Landscape Complexity

Ecosystem Dynamics, Economic Pressure and Optimal Resource Management

RIVER PUBLISHERS SERIES IN SOCIAL, URBAN, ECONOMIC AND ENVIRONMENTAL SUSTAINABILITY

The River Series on Social, Urban, Economic and Environmental Sustainability is a series of comprehensive academic and professional books which focus on the societal side of sustainability. The series focuses on topics ranging from theory to policy and real-life case studies and applications.

Books published in the series include research monographs, edited volumes, handbooks and textbooks. The books provide professionals, researchers, educators, and advanced students in the field with an invaluable insight into the latest research and developments.

Topics included in this series are as follows:-

- Climate Change Mitigation
- Renewable Energy Policy
- Urban sustainability
- Strategic environmental planning
- Environmental Systems Monitoring and Analysis
- Greening the World Economy
- Sustainable Development

For a list of other books in this series, visit www.riverpublishers.com

Landscape Complexity

Ecosystem Dynamics, Economic Pressure and Optimal Resource Management

Editors

Alessia D'Agata
Autonomous University of Barcelona,
Spain

Ioannis Konaxis
University of Piraeus,
Greece

Luca Salvati
Sapienza University of Rome,
Italy

River Publishers

Routledge
Taylor & Francis Group
NEW YORK AND LONDON

Published 2026 by River Publishers
River Publishers
Broagervej 10, 9260 Gistrup, Denmark
www.riverpublishers.com

Distributed exclusively by Routledge
605 Third Avenue, New York, NY 10017, USA
4 Park Square, Milton Park, Abingdon, Oxon OX14 4RN

Landscape Complexity: Ecosystem Dynamics, Economic Pressure and Optimal Resource Management / Alessia D'Agata, Ioannis Konaxis and Luca Salvati.

Routledge is an imprint of the Taylor & Francis Group, an informa business

ISBN 978-87-438-1108-4 (hardback)
ISBN 978-87-438-1114-5 (paperback)
ISBN 978-87-438-1111-4 (online)
ISBN 978-87-438-1113-8 (e-pub)
ISBN 978-8-743-81110-7 (ebook master)
DOI 10.1201/9788743811107

Contents

Acknowledgment

This study was partially supported by the MULTIFOR "Multi-scale observations to predict Forest response to pollution and climate change" PRIN 2020 Research Project of National Relevance funded by the Italian Ministry of University and Research (prot. 2020E52THS).

List of Contributors

Bianchini, Roberto, *Department of Agricultural and Forest Sciences (DAFNE), University of Tuscia, Italy*

Ciaschini, Clio, *Department of Economics and Social Sciences, Polytechnic University of Marche, Italy*

Colantoni, Andrea, *Department of Agricultural and Forest Sciences (DAFNE), University of Tuscia, Italy*

D'Agata, Alessia, *Department of Methods and Models for Economics, Territory and Finance (MEMOTEF), Faculty of Economics, Sapienza University of Rome, Italy; Department of Geography, Building B, Autonomous University of Barcelona, Spain*

Egidi, Gianluca, *Department of Agricultural and Forest Sciences (DAFNE), University of Tuscia, Italy*

Imbrenda, Vito, *Institute of Methodologies for Environmental Analysis, Italian National Research Council (CNR-IMAA), Italy*

Konaxis, Ioannis, *Department of Tourism Studies, University of Piraeus, Greece*

Lombardi, Pierpaolo, *Gruppo Maurizi, Italy*

Maialetti, Marco, *Independent Researcher and Lawyer, Italy*

Mavrakis, Anastasios, *Secondary Education Directorate of West Attica, Greek Ministry of Education, Greece*

Mosconi, Enrico Maria, *Department of Economics, Engineering, Society and Business Organization (DEIM), University of Tuscia, Italy*

Nosova, Bogdana, *Department of Social Communications, Institute of Journalism, Taras Shevchenko National University of Kyiv, Ukraine*

Papakitsos, Evangelos, *Geo-Environmental Science and Environmental Quality Assurance Laboratory, University of West Attica, Greece*

Papavasileiou, Christina, *Second Athens Secondary Education Directorate, Greek Ministry of Education, Greece*

Quaranta, Giovanni, *Department of Mathematics, Informatics and Economics, University of Basilicata, Italy*

Salvati, Luca, *Department of Methods and Models for Economics, Territory and Finance (MEMOTEF), Faculty of Economics, Sapienza University of Rome, Italy*

Salvia, Rosanna, *Department of Mathematics, Informatics and Economics, University of Basilicata, Italy*

Salvucci, Gianluigi, *Italian National Institute of Statistics (ISTAT), Italy*

Sateriano, Adele, *Mediterranean Foundation for Sustainable Development (MEDES), Italy*

Scarpitta, Donato, *Independent Researcher and Architect, Italy*

Verouti, Eleni, *Head of the Administration of Environment & Civil Protection at Municipality of Aspropyrgos, Greece*

List of Figures and Table

List of Abbreviations

COPI	Conference of the Parties
EE	Ecological Education
ESA	Environmentally Sensitive Areas
EU	European Union
GDP	Gross Domestic Product
GIS	Geographical information system
GWR	Geographically weighted regression
HEI	Human-environmental interaction
IPCC	Intergovernmental Panel on Climate Change
KMO	Kaiser–Meyer–Olkin
MEA	Millennium Ecosystem Assessment
OECD	Organization for Economic Co-operation and Development
PCA	Principal component analysis
SDG	Sustainable Development Goal
SDM	Spatial Durbin model
UA	Urban Atlas
UNCCD	United Nations Convention to Combat Desertification
UNFCCC	United Nations Framework Convention on Climate Change

1

Socio-ecological Assessment of Landscape Systems: Toward an Integrated Perspective

Alessia D'Agata[1,2], Clio Ciaschini[3], Marco Maialetti[4], Donato Scarpitta[5], Gianluca Egidi[6], Roberto Bianchini[6], and Ioannis Konaxis[7]

[1]Department of Methods and Models for Economics, Territory and Finance (MEMOTEF), Faculty of Economics, Sapienza University of Rome, Italy
[2]Department of Geography, Building B, Autonomous University of Barcelona, Spain
[3]Department of Economics and Social Sciences, Polytechnic University of Marche, Italy
[4]Independent Researcher and Lawyer, Italy
[5]Independent Researcher and Architect, Italy
[6]Department of Agricultural and Forest Sciences (DAFNE), University of Tuscia, Italy
[7]Department of Tourism Studies, University of Piraeus, Greece

Email: alessia.dagata@uniroma1.it; c.ciaschini@univpm.it; avv.marcomaialetti@gmail.com; scarpitta.donato@libero.it; egidi.gianluca@unitus.it; roberto.bianchini@unitus.it; ikonaxis@unipi.gr

Abstract

Socio-ecological and landscape models typically reflect self-organized development rather than patterns controlled by a central institution, even in the face of exogenous stresses (e.g., climate change) impacting their sustainability characteristics and resilience potential. Based on this assumption, the present contribution discusses complex landscape systems from a multidisciplinary perspective, providing concrete environmental and economic examples, and revealing – from an operational and planning perspective – how "complex thinking and acting" is useful not only for landscape science but also for policy implications.

DOI: 10.1201/9788743811107-1

1.1 Introduction

Forests and rural habitats have played a crucial role in the rise and development of some of the oldest civilizations, thanks to their multi-functional role, which offered a wide range of benefits and services to society and human welfare beyond traditional forest wood products (Wilson et al., 2017). In the long term, a mosaic-like structure emerged from the complex interactions of forests and shrubland with other land use types, such as agriculture, pasture, and urban/industrial areas (Ferrara et al., 2016). This leads to complexity levels that directly influence landscapes in a joint ecological and socioeconomic perspective (Ferrara et al., 2020). Despite the emergence of specific tools and programs for landscape evaluation, the lack of a national accounting system for landscape analysis – can be partly responsible for the limited recognition of the relevant role of landscapes in economic systems (Salvati et al., 2015). However, it is widely recognized that forest services provide more benefits to the local communities than to their individual owners, making it even more difficult to evaluate their contribution (Ferrara et al., 2017).

The basic contribution of this book is grounded on the assumption that official statistics, field surveys, remote monitoring, landscape analysis, and forest inventories, among others, can play an important role in assessing the quantitative (and most of the qualitative) aspects of ecosystem services (Colantoni et al., 2015). These include "ecological support" (i.e., biodiversity conservation, wildlife protection, water supply and water quality, soil fertility recovery/generation, and employment support), "provision" (i.e., timber products, non-wood forest products, renewable energy, and food, pasture), and "regulation" (i.e., climate and microclimate regulation, carbon sequestration, water regulation, drought mitigation, and soil erosion protection/prevention). Such methodologies, however, are also assumed to shape some of the most important cultural services that they can provide, namely cultural-natural heritage, inspiration and aesthetic values, or recreational assets and ecotourism quality (Tomao et al., 2017).

A basic concept of the literature involves the notion of "carrying capacity" characteristic of a given territory and its community (Coluzzi et al., 2022). Focusing on both agricultural systems and natural land-use, the proposed notion of "carrying capacity" is intended to evaluate holistically the potential of any territory to reflect changes in both environmental and socioeconomic issues, and answer future challenges, stemming from internal changes in the agricultural sector, as well as ecological modifications (e.g., climate change), societal transitions, and (expected or unexpected) economic dynamics (Salvati et al., 2015, 2017).

To fulfil this main and ambitious objective, of particular interest for policy implementation, three dimensions have been considered here, representing (i) future climate conditions (e.g., soil/climate aridity) (Colantoni et al., 2015), (ii) demography, mainly capturing rural population dynamics, in connection with broader demographic dynamics at both urban and country scale, as a proxy of socioeconomic dynamics at large (Imbrenda et al., 2021), and (iii) the intrinsic capacity of the territories to invest in their future agriculture (Quaranta et al., 2020). Gross Domestic Product (GDP) is a proxy indicator for this dimension, as a general measure of responsiveness of the economic system to address future issues in the agricultural field – from the ecological crisis to depopulation, from market uncertainties to modifications in consumers' preferences and lifestyles (Salvati et al., 2015).

This indicator additionally provides better insight into how future policies can territorially address territorially specific land use trends in the context of their carrying capacity asset (Coluzzi et al., 2022). Specific policy recommendations can be formulated taking account of the combined characteristics of land-use classes and regions/districts (Ferrara et al., 2017). This will require serious efforts in terms of policy implementation, and if no active policy interventions are in place, the decline of natural areas and farmlands is expected to be significant. This will be accompanied by a large increase in urban land take of agricultural land if no policy measures are implemented to prevent such trends (Salvati et al., 2013). In such contexts, land abandonment can also expand following depopulation in marginal areas, worse climate conditions for farming, and an unwanted expansion of non-managed (low-quality and biodiversity-poor) forests – possibly out of the economic value chain because of an intrinsic lack of management and limited ecosystem values.

1.2 Positive and Normative Approaches for Complex Landscape Systems

Thanks to accelerated economic growth, agro-environmental degradation has emerged as a global issue, particularly given agriculture's dual role as both a driver and victim of environmental change (Ferrara et al., 2016). Unsustainable farming practices, land-use transformations, and socioeconomic pressures have accelerated soil degradation, leading to compaction, erosion, nutrient depletion, and biodiversity loss (Quaranta et al., 2020). Since the mid-20th century, processes of agricultural specialization, urban expansion, and rural abandonment have reshaped landscapes, exacerbating conflicts over natural resources and intensifying pressures on soil and water systems (Coluzzi et al., 2022). Key manifestations of landscape degradation include pollution from pesticides and fertilizers, desertification, and the homogenization of agro-ecosystems with the spread of dominant crops (Ferrara et al., 2020).

Moreover, climate change – closely associated with land-use practices – has amplified such threats, contributing to local warming, altered hydrological cycles, and increased risks of extreme events, with profound implications for individual landscapes and regions (Ferrara et al., 2017). Soil degradation and early desertification processes – recognized as major environmental priorities by the international community – further threaten ecosystem services, food security, and human well-being (Salvati et al., 2015). Habitat fragmentation and land consumption compound these dynamics by accelerating biodiversity loss and reducing ecological resilience (Coluzzi et al., 2022). Based on this rationale, Chapter 2 underscores the urgent need for integrated strategies that address the interplay between agriculture, land management, and climate change to mitigate agro-environmental degradation and safeguard socio-ecological systems (Ferrara et al., 2016).

Taken as another environmental issue with basically complex socio-economic implications, global (and local) warming reflects a profound and interlinked challenge for both natural ecosystems and human societies, necessitating prompt and coordinated mitigation and adaptation strategies (Ferrara et al., 2017). Understanding and managing these dynamics requires multidimensional approaches capable of integrating environmental, economic, and social knowledge (Wilson et al., 2017). Current methodologies for quantifying ecosystem states, including natural capital and ecosystem services, provide essential information but often fail to capture the resilience of local communities or the complex interactions between human and environmental systems (Ferrara et al., 2020). The contribution emphasizes the development of composite, locally relevant indicators that bridge disciplinary boundaries, combining official statistics, geospatial data, and remote sensing with qualitative and historical knowledge (Salvati et al., 2017). Such indicators are designed to inform sustainable land management, strategic planning, and policy decisions aimed at mitigating climate impacts, particularly in forests and natural habitats (Ferrara et al., 2016).

By linking socioeconomic patterns, demographic dynamics, and environmental changes, this approach contributes to a more nuanced understanding of spatial inequalities, ecosystem functionality, and the effects of human activities on landscape services. Multidisciplinary methodologies, including spatial econometrics, remote sensing, and integrated statistical analysis, allow for a comprehensive assessment of land-use change, forest cover dynamics, and the socio-environmental determinants of ecosystem health (Salvati et al., 2017). Ultimately, this approach supports evidence-based policy-making, facilitating adaptation and mitigation strategies aligned with sustainable development goals, while highlighting potential inequalities in resource access and environmental benefits (Ferrara et al., 2020). The operational framework proposed here emphasizes the importance of local-scale, holistic, and temporally informed indicators as a critical foundation for resilient and equitable ecosystem governance in the context of climate change (Wilson et al., 2017).

1.2.1 System complexity and "holistic" landscape science

Traditional land-use practices have historically shaped resilient socio-ecological systems. Among the world's most significant biodiversity hotspots, terraced landscapes represent complex socio-ecological systems that stabilize steep slopes, support agricultural production, and provide diversified ecosystem services while reflecting centuries of accumulated local knowledge (Ferrara et al., 2017). Despite their cultural, ecological, and economic value, terraced landscapes are threatened by abandonment, degradation, and socio-economic marginalization, compounded by climate change and rural shrinkage (Salvati et al., 2015). Sustainable management of these systems requires a comprehensive understanding of their ecological functions, socioeconomic dimensions, and historical transformations (Ferrara et al., 2020). Integrative monitoring approaches, combining multi-temporal datasets, remote sensing, field surveys, and geographic information systems, enable the evaluation of biodiversity, ecosystem services, land-use dynamics, and traditional ecological knowledge (Salvati et al., 2017).

Such assessments can inform strategies for terrace stabilization, restoration of degraded areas, and adaptation of traditional practices to contemporary socio-environmental conditions (Ferrara et al., 2016). Holistic analysis of terraced landscapes emphasizes their multifunctional role in promoting biodiversity conservation, supporting agro-forestry and multifunctional agriculture, and fostering socio-cultural and economic resilience in rural areas (Figure 1.1). By integrating ecological, agricultural, and socioeconomic perspectives, these frameworks facilitate informed decision-making, landscape planning, and policy development (Wilson et al., 2017). Preserving and revitalizing terraced systems not only safeguards cultural heritage but also enhances ecological stability, supports local livelihoods, and contributes to sustainable rural development in a context of global environmental change (Ferrara et al., 2020).

Figure 1.1 Complex landscape structures, aesthetic values, and biodiversity in a coastal Mediterranean pinewood under legal protection as a regional park; source: authors' photographic archive.

Moving on to a purely quantitative perspective, Chapter 5 introduces a multidimensional framework for examining the composition of metropolitan landscapes through an extended form of principal component analysis (PCA) that incorporates alternative correlation metrics (Ferrara et al., 2016). Traditional PCA remains one of the most widely applied techniques in exploratory data analysis and predictive modeling, yet its reliance on linear relationships and its sensitivity to outliers impose significant constraints when applied to complex spatial data (Salvati et al., 2017). Metropolitan land-use systems are rarely governed by strictly linear associations, and the presence of outliers is often intrinsic to heterogeneous urban contexts (Ferrara et al., 2020). These limitations underscore the need for alternative approaches capable of revealing the latent structures embedded in large and multifaceted datasets (Wilson et al., 2017).

To address these methodological challenges, three PCA implementations were compared using Pearson, Spearman, and Kendall correlation matrices. The empirical case study involved land-use data from local administrative units within metropolitan Athens, Greece, derived from the Urban Atlas initiative (Ferrara et al., 2017). The dataset describes the proportional distribution of several land-use categories, thus offering a high-resolution representation of the spatial heterogeneity of metropolitan landscapes (Salvati et al., 2017). By applying different correlation metrics, the study sought to determine how alternative conceptualizations of latent associations among land-use types influence dimensionality reduction and the interpretability of principal components (Ferrara et al., 2020). The results demonstrate that PCA based on Spearman's rank correlation provided a superior analytical performance in multiple respects.

While Pearson's correlation accounted for the greatest share of variance in the first component, it proved less effective in subsequent dimensions, where Spearman consistently offered clearer discrimination among cases and more stable component loadings. The diagnostic indicators further reinforced this observation: both the Kaiser–Meyer–Olkin test and Bartlett's test suggested higher adequacy and stronger suitability for the Spearman-based solution (Ferrara et al., 2017). The robustness of Spearman's approach lies in its non-parametric nature, which makes it less vulnerable to the distortions caused by outliers and more capable of capturing monotonic, potentially non-linear relationships that characterize spatial and environmental data. Kendall's tau, while valuable in specific contexts, showed a more conservative explanatory capacity, highlighting that the choice of correlation metric directly affects the quality of dimensionality reduction.

1.3 Landscape Complexity and Planning Implications

In a context of accelerated economic growth and rising human pressure on natural landscapes, wildfires are becoming increasingly dangerous and geographically widespread in advanced countries, harming environmental quality and negatively impacting both the structure and functions of ecosystems (Ferrara et al., 2016). In such a context, fire prevention strategies have to deal with evolving environmental and socioeconomic scenarios (Ferrara et al., 2020). The combination of a warming climate, fuel patterns, and socioeconomic conditions is responsible for increasing large fire occurrences throughout Europe (Wilson et al., 2017). In the last decades, simultaneous large fires often overwhelmed the suppression capacity of single countries – not only in fire-prone areas, but also in countries occasionally affected by such events (Ferrara et al., 2017). In the latter countries, wildfires resulted in severe and long-lasting effects due to the lower fire resistance and resilience of forests. In order to cope with a changing landscape increasingly affected by unpredictable fires, Chapter 6 focuses on fire management strategies and discusses prevention actions based on broader sustainable land/forest management aimed at increasing response preparedness. These objectives will be reached by sharing national, regional, and local (good) practices among countries with different operational experiences and skills, underlying different levels of response preparedness.

Corroborating the positive impact of sustainable land planning, Ecological Education (EE) emerged as a core component of environmental policy in response to the growing concerns about ecosystem degradation, pollution, and the depletion of natural resources (Ferrara et al., 2016). In this vein, Chapter 7 illustrates the pedagogical methodology employed, the students' responses and engagement, and the overall experience of the participating educators of an environmental education project entitled "Koumoundourou lake: a Habitat in our yard," integrated into a broader thematic context of "Human rights in a clean environment and culture." The initiative offered students an opportunity to explore a nearby yet often overlooked wetland ecosystem of both ecological and historical importance (Salvati et al., 2017).

Through active participation, research, and fieldwork, students developed critical environmental awareness, historical insight, and a personal connection to their local landscape (Ferrara et al., 2016). With the purpose of making this particular topic of environmental education programs more attractive, the use of relevant web tools has been additionally discussed (Ferrara et al., 2017). The empirical results of this study suggest how modernizing the educational process, and especially upgrading the web content of schools' programs, is particularly urgent (Wilson et al., 2017).

When pursued over extended temporal horizons, the systematic examination of demographic dynamics finally offers a privileged perspective to envision complex socioeconomic processes underlying the genesis and persistence of territorial disparities and inequalities at the landscape scale – whether within regions, across nations, or at continental scales (Cassiers & Kesteloot, 2012). Nevertheless, notwithstanding the centrality of such approaches, empirical inquiries into the long-term role of population density in structuring human settlement systems along urban–rural gradients remain comparatively limited, even in world areas marked by an exceptionally deep settlement history, such as Europe and the Mediterranean basin (Ferrara et al., 2016).

In recent years, selected investigations have endeavored to disentangle these dynamics through a set of case studies ranging from the global scale to the European and national levels, making use of advanced data mining procedures and spatial econometric models – both cross-sectional and longitudinal – in their global, quantile, and local formulations (Arapoglou & Sayas, 2009). As a contribution to planning the socioeconomic complexity inherent in each landscape, Chapter 8 discusses the extent to which density-dependent and path-dependent mechanisms correspond to specific phases within broader economic/building/metropolitan cycles (Bruegmann, 2005).

In addition, the contribution foregrounds the intricate interplay between economic forces – such as agglomeration dynamics and economies of scale – and non-economic dimensions, including institutional arrangements, cultural specificities, and territorial structures, all situated within a historically informed regional perspective (Wilson et al., 2017). Through the interpretation of spatio-temporal economic configurations and socio-demographic indicators, recent studies ultimately aspire to inform the design of regional development strategies and spatial planning policies capable of responding flexibly to heterogeneous and rapidly evolving local contexts.

1.4 Concluding Remarks

Rural spaces are significantly differentiated as far as land-use, economic dynamics, social structures, and ecological conditions are concerned. All these conditions intrinsically influence agricultural and forest dynamics and drive territories to respond to policy stimuli in a possibly different way. We documented such disparities adopting a multi-thematic and interdisciplinary perspective using a kaleidoscope of different visions, approaches, methodologies, analysis techniques, and indicators. In this perspective, the notion of "carrying capacity" is assumed to be particularly appropriate to evaluate – from a territorial perspective – the different states and ability of landscape systems to respond to policy stimuli, based on the assets of conditions and contexts available at the moment and possibly influencing the individual ability of landscape systems to exploit policy opportunities.

References

Arapoglou, V. P., & Sayas, J. (2009). New facets of urban segregation in southern Europe. European Urban and Regional Studies, 16(4), 345–362.

Batty, M., & Longley, P. (1994). Fractal cities. London: Academic Press.

Beriatos, E., & Gospodini, A. (2004). "Glocalising" urban landscapes: Athens and the 2004 Olympics. Cities, 21, 187–202.

Berry, B. J. L. (2005). Cities as systems within systems of cities. Papers in Regional Science, 13, 147–163.

Brueckner, J. K., & Fansler, D. (1983). The economics of urban sprawl: Theory and evidence on the spatial sizes of cities. Review of Economics and Statistics, 55, 479–482.

Bruegmann, R. (2005). Sprawl: A compact history. Chicago: University of Chicago Press.

Cassiers, T., & Kesteloot, C. (2012). Socio-spatial inequalities and social cohesion in European cities. Urban Studies, 49(9), 1909–1924.

Catalán, B., Saurí, D., & Serra, P. (2008). Urban sprawl in the Mediterranean? Patterns of growth and change in the Barcelona Metropolitan Region 1993–2000. Landscape and Urban Planning, 85(3–4), 174–184.

Chorianopoulos, I. (2003). North-south local authority and governance differences in the EU networks. European Planning Studies, 11(6), 671–695.

Chorianopoulos, I., Pagonis, T., Koukoulas, S., & Drymoniti, S. (2010). Planning, competitiveness and sprawl in the Mediterranean city: The case of Athens. Cities, 27, 249–259.

Coccossis, H., Economou, D., & Petrakos, G. (2005). The ESDP relevance to a distant partner: Greece. European Planning Studies, 13(2), 253–264.

Colantoni, A., Ferrara, C., Perini, L., & Salvati, L. (2015). Assessing trends in climate aridity and vulnerability to soil degradation in Italy. Ecological Indicators, 48, 599–604.

Coluzzi, R., Bianchini, L., Egidi, G., Cudlin, P., Imbrenda, V., Salvati, L., & Lanfredi, M. (2022). Density matters? Settlement expansion and land degradation in peri-urban and rural districts of Italy. Environmental Impact Assessment Review, 92, 106703.

Costa, F., Noble, A. G., & Pendleton, G. (1991). Evolving planning systems in Madrid, Rome and Athens. Geojournal, 24, 293–303.

Couch, C., Petschel-Held, G., & Leontidou, L. (2007). Urban sprawl in Europe: Landscapes, land-use change and policy. London: Blackwell.

Delladetsima, P. (2006). The emerging property development pattern in Greece and its impact on spatial development. European Urban and Regional Studies, 13(3), 245–278.

Domènech, E., & Saurí, D. (2006). Urbanisation and water consumption: Influencing factors in the metropolitan region of Barcelona. Urban Studies, 43(9), 1605–1623.

Durà-Guimera, A. (2003). Population deconcentration and social restructuring in Barcelona, a European Mediterranean city. Cities, 20(6), 387–394.

Economidou, E. (1993). The Attic landscape throughout the centuries and its human degradation. Landscape and Urban Planning, 24, 33–37.

Economou, D. (1997). The planning system and rural land use control in Greece: A European perspective. European Planning Studies, 5(4), 461–476.

Economou, D., Petrakos, G., & Psycharis, Y. (2007). Urban policy in Greece. In L. Van den Berg, E. Braun, & J. Van der Meer (Eds.), National policy responses to urban challenges in Europe (pp. xx–xx). Ashgate.

Egidi, G., Salvati, L., & Vinci, S. (2020). The long way to Tipperary: City size and worldwide urban population trends, 1950–2030. Sustainable Cities and Society, 60, 102148.

Ezcurra, R., & Rapun, M. (2006). Regional disparities and national development revisited: The case of western Europe. European Urban and Regional Studies, 13, 355–369.

Ferrara, A., Kelly, C., Wilson, G. A., Nolè, A., Mancino, G., Bajocco, S., & Salvati, L. (2016). Shaping the role of 'fast' and 'slow' drivers of change in forest-shrubland socio-ecological systems. Journal of Environmental Management, 169, 155–166.

Ferrara, A., Kosmas, C., Salvati, L., Padula, A., Mancino, G., & Nolè, A. (2020). Updating the MEDALUS ESA framework for worldwide land degradation and desertification assessment. Land Degradation & Development, 31(12), 1593–1607.

Ferrara, C., Carlucci, M., Grigoriadis, E., Corona, P., & Salvati, L. (2017). A comprehensive insight into the geography of forest cover in Italy: Exploring the importance of socioeconomic local contexts. Forest Policy and Economics, 75, 12–22.

Francaviglia, R., Di Bene, C., Farina, R., Salvati, L., & Vicente-Vicente, J. L. (2019). Assessing "4 per 1000" soil organic carbon storage rates under Mediterranean climate: A comprehensive data analysis. Mitigation and Adaptation Strategies for Global Change, 24, 795–818.

Frenkel, A., & Ashkenazi, M. (2008). Measuring urban sprawl: How can we deal with it? Environment and Planning B: Planning and Design, 35, 56–79.

Garcia-López, M. A., & Muniz, I. (2010). Employment decentralisation: Polycentricity or scatteration? The case of Barcelona. Urban Studies, 47(14), 3035–3056.

Giaccaria, P., & Minca, C. (2010). The Mediterranean alternative. Progress in Human Geography, 35(3), 345–365.

Glaster, G., Hanson, R., Ratcliffe, M. R., Wolman, H., Coleman, S., & Freihage, J. (2001). Wrestling sprawl to the ground: Defining and measuring an elusive concept. Housing Policy Debate, 12(4), 681–717.

Gospodini, A. (2001). Urban waterfront redevelopment in Greek cities. Cities, 18(5), 285–295.

Gospodini, A. (2006). Portraying, classifying and understanding the emerging landscapes in the post-industrial city. Cities, 23(5), 311–330.

Gospodini, A. (2009). Post-industrial trajectories of Mediterranean European cities: The case of post-Olympics Athens. Urban Studies, 46(5–6), 1157–1186.

Holland, J. H. (2006). Studying complex adaptive systems. Journal of Systems Science and Complexity, 19(1), 1–8.

Imbrenda, V., Quaranta, G., Salvia, R., Egidi, G., Salvati, L., Prokopovà, M., Coluzzi, R., & Lanfredi, M. (2021). Land degradation and metropolitan expansion in a peri-urban environment. Geomatics, Natural Hazards and Risk, 12(1), 1797–1818.

Incerti, G., Feoli, E., Salvati, L., Brunetti, A., & Giovacchini, A. (2007). Analysis of bioclimatic time series and their neural network-based classification to characterise drought risk patterns in South Italy. International Journal of Biometeorology, 51, 253–263.

Kazepov, Y. (2005). Cities of Europe: Changing contexts, local arrangements, and the challenge to urban cohesion. Blackwell, London.

Keyder, C. (2005). Globalization and social exclusion in Istanbul. International Journal of Urban and Regional Research, 29(1), 124–134.

King, R., Proudfoot, L., & Smith, B. (1997). The Mediterranean: Environment and society. Arnold.

Kosmas, C., Detsis, V., Karamesouti, M., Kounalaki, K., Vassiliou, P., & Salvati, L. (2015). Exploring long-term impact of grazing management on land degradation in the socio-ecological system of Asteroussia Mountains, Greece. Land, 4(3), 541–559.

Kourliouros, E. (1997). Planning industrial location in Greater Athens: The interaction between deindustrialization and anti-industrialism during the 1980s. European Planning Studies, 5(4), 435–460.

Kourliouros, E. (2003). Reflections on the economic-noneconomic debate: A radical geographical perspective from the European south. Antipode, 35(4), 781–799.

Kourtit, K., Nijkamp, P., & Reid, N. (2014). The new urban world: Challenges and policy. Applied Geography, 49, 1–3.

Kresl, P. K. (2007). Planning cities for the future: The successes and failures of urban economic strategies in Europe. Edward Elgar.

Krumholz, N. (1992). Roman impressions: Contemporary city planning and housing in Rome. Landscape and Urban Planning, 22(2–4), 107–114.

Lambooy, J. G. (1998). Polynucleation and economic development: The Randstad. European Planning Studies, 6(4), 457–467.

Lamonica, G. R., Recchioni, M. C., Chelli, F. M., & Salvati, L. (2020). The efficiency of the cross-entropy method when estimating the technical coefficients of input–output tables. Spatial Economic Analysis, 15(1), 62–91.

Leontidou, L. (1990). The Mediterranean city in transition. Cambridge University Press.

Leontidou, L. (1993). Postmodernism and the city: Mediterranean versions. Urban Studies, 30, 949–965.

Leontidou, L. (1996). Alternatives to modernism in (Southern) urban theory: Exploring in-between spaces. International Journal of Urban and Regional Research, 20(2), 180–197.

Leontidou, L. (2012). Athens in the Mediterranean 'movement of the piazzas'. Spontaneity in material and virtual spaces. City: Analysis of Urban Trends, Culture, Theory, Policy, Action, 16(3), 299–312.

Longhi, C., & Musolesi, A. (2007). European cities in the process of economic integration: Towards structural convergence. Annals of Regional Science, 41, 333–351.

López-i-Villanueva, C., Pujadas, R. I., & Carrasco, J. B. (2013). Households within the residential mobility process: The case of the Barcelona metropolitan region. Archivio di Studi Urbani e Regionali, 28, 57–84.

Maloutas, T. (2007a). Segregation, social polarization and immigration in Athens during the 1990s: Theoretical expectations and contextual difference. International Journal of Urban and Regional Research, 31(4), 733–758.

Maloutas, T. (2007b). Socio-economic classification models and contextual difference: The "European socio-economic classes" (ESeC) from a South European angle. South European Society & Politics, 12(4), 443–460.

Maloutas, T., & Karadimitriou, N. (2001). Vertical social differentiation in Athens: Alternative or complement to community segregation. International Journal of Urban and Regional Research, 25(4), 699–716.

Mancino, G., Nolè, A., Salvati, L., & Ferrara, A. (2016). In-between forest expansion and cropland decline: A revised USLE model for soil erosion risk under land-use change in a Mediterranean region. Ecological Indicators, 71, 544–550.

McNeill, D. (1999). Globalization and the European city. Cities, 16(3), 143–148.

Minca, C. (2003). Critical peripheries. Environment and Planning D: Society and Space, 21, 160–168.

Mudu, P. (2006). Patterns of segregation in contemporary Rome. Urban Geography, 27(5), 422–440.

Munafò, M., Norero, C., Sabbi, A., & Salvati, L. (2010). Soil sealing in the growing city: A survey in Rome, Italy. Scottish Geographical Journal, 126(3), 153–161.

Munõz, F. (2003). Lock living: Urban sprawl in Mediterranean cities. Cities, 20(6), 381–385.

Musterd, S., & Ostendorf, W. (1998). Urban segregation and the welfare state: Inequality and exclusion in western cities. Routledge.

Orenstein, D. E., Frenkel, A., & Jahshan, F. (2014). Methodology matters: Measuring urban spatial development using alternative methods. Environment and Planning B: Planning and Design, 41, 3–23.

Pacione, M. (2003). Urban environmental quality and human wellbeing—a social geographical perspective. Landscape and Urban Planning, 65, 19–30.

Phelps, A., Parsons, A., Ballas, D., & Dowling, A. (2006). Post-suburban Europe: Planning and politics at the margins of Europe's capital cities. Palgrave Macmillan.

Portnov, B. A., & Safriel, U. N. (2004). Combating desertification in the Negev: Dryland agriculture vs. dryland urbanization. Journal of Arid Environments, 56, 659–680.

Quaranta, G., Salvia, R., Salvati, L., Paola, V. D., Coluzzi, R., Imbrenda, V., & Simoniello, T. (2020). Long-term impacts of grazing management on land degradation in a rural community of Southern Italy: Depopulation matters. Land Degradation & Development, 31(16), 2379–2394.

Sallustio, L., Pettenella, D., Merlini, P., Romano, R., Salvati, L., Marchetti, M., & Corona, P. (2018). Assessing the economic marginality of agricultural lands in Italy to support land use planning. Land Use Policy, 76, 526–534.

Salvati, L., & De Rosa, S. (2014). Hidden polycentrism or subtle dispersion? Urban growth and long-term sub-centre dynamics in three Mediterranean cities. Land Use Policy, 39, 233–243.

Salvati, L., Zitti, M., & Sateriano, A. (2013). Changes in the city vertical profile as an indicator of sprawl: Evidence from a Mediterranean region. Habitat International, 38, 119–125.

Salvati, L. (2013). Monitoring high-quality soil consumption driven by urban pressure in a growing city (Rome, Italy). Cities, 31, 349–356.

Salvati, L., & Carlucci, M. (2011). The economic and environmental performances of rural districts in Italy: Are competitiveness and sustainability compatible targets? Ecological Economics, 70(12), 2446–2453.

Salvati, L., & Zitti, M. (2008). Regional convergence of environmental variables: Empirical evidences from land degradation. Ecological Economics, 68(1–2), 162–168.

Salvati, L., & Zitti, M. (2009). Substitutability and weighting of ecological and economic indicators: Exploring the importance of various components of a synthetic index. Ecological Economics, 68(4), 1093–1099.

Salvati, L., Mavrakis, A., Colantoni, A., Mancino, G., & Ferrara, A. (2015). Complex adaptive systems, soil degradation and land sensitivity to desertification: A multivariate assessment of Italian agro-forest landscape. Science of the Total Environment, 521, 235–245.

Salvati, L., Perini, L., Sabbi, A., & Bajocco, S. (2012). Climate aridity and land use changes: A regional-scale analysis. Geographical Research, 50(2), 193–203.

Salvati, L., Ranalli, F., Carlucci, M., Ippolito, A., Ferrara, A., & Corona, P. (2017). Forest and the city: A multivariate analysis of peri-urban forest land cover patterns in 283 European metropolitan areas. Ecological Indicators, 73, 369–377.

Salvati, L., Tombolini, I., Perini, L., & Ferrara, A. (2013). Landscape changes and environmental quality: The evolution of land vulnerability and potential resilience to degradation in Italy. Regional Environmental Change, 13, 1223–1233.

Salvati, L., Venanzoni, G., Serra, P., & Carlucci, M. (2016). Scattered or polycentric? Untangling urban growth in three southern European metropolitan regions through exploratory spatial data analysis. The Annals of Regional Science, 57, 1–29.

Saurí, D. (2003). Lights and shadows of urban water demand management: The case of the metropolitan region of Barcelona. European Planning Studies, 11(3), 229–243.

Schneider, A., & Woodcock, C. E. (2008). Compact, dispersed, fragmented, extensive? A comparison of urban growth in twenty-five global cities using remotely sensed data, pattern metrics and census information. Urban Studies, 45(3), 659–692.

Scott, A. J. (2001). Globalization and the rise of city-regions. European Planning Studies, 9(7), 813–826.

Scott, A. J. (2001). Global city-regions: Trends, theory, policy. Oxford University Press, Oxford.

Seifollahi-Aghmiuni, S., Kalantari, Z., Egidi, G., Gaburova, L., & Salvati, L. (2022). Urbanisation-driven land degradation and socioeconomic challenges in peri-urban areas: Insights from Southern Europe. Ambio, 51(6), 1446–1458.

Souliotis, N. (2013). Cultural economy, sovereign debt crisis and the importance of local contexts: The case of Athens. Cities, 33, 61–68.

Tomao, A., Quatrini, V., Corona, P., Ferrara, A., Lafortezza, R., & Salvati, L. (2017). Resilient landscapes in Mediterranean urban areas: Understanding factors influencing forest trends. Environmental Research, 156, 1–9.

Torrens, P. M. (2008). A toolkit for measuring sprawl. Applied Spatial Analysis and Policy, 1, 5–36.

Tsai, Y. H. (2005). Quantifying urban form: Compactness versus ‘sprawl’. Urban Studies, 42(1), 141–161.

Tsilimpounidi, M. (2012). Athens 2012. City: Analysis of urban trends, culture, theory, policy, action, 16(5), 546–556.

Turok, I., & Mykhnenko, V. (2007). The trajectories of European cities, 1960–2005. Cities, 24(3), 165–182.

Vaiou, D. (1997). Facets of spatial development and planning in Greece. European Planning Studies, 5(4), 431–433.

Vardopoulos, I., Tsilika, E., Sarantakou, E., Zorpas, A. A., Salvati, L., & Tsartas, P. (2021). An integrated SWOT-PESTLE-AHP model assessing sustainability in adaptive reuse projects. Applied Sciences, 11(15), 7134.

Vidal, M., Domènech, E., & Saurí, D. (2011). Changing geographies of water-related consumption: Residential swimming pools in suburban Barcelona. Area, 43(1), 67–75.

Williams, A. M. (1984). Southern Europe transformed: Political and economic change in Greece, Italy, Portugal and Spain. London: Harper and Row.

Wilson, G. A., Kelly, C. L., Briassoulis, H., Ferrara, A., Quaranta, G., Salvia, R., Detsis, V., Curfs, M., Cerda, A., El-Aich, A., Liu, H., Kosmas, C., Alados, C. L., Imeson, A., Landgrebe-Trinkunaite, R., Salvati, L., Naumann, S., Danwen, H., Iosifides, T., Kizos, T., Mancino, G., & Nolè, A. (2017). Social memory and the resilience of communities affected by land degradation. Land Degradation & Development, 28(2), 383–400.

2

Environmental Degradation in Complex Landscapes: From Theory to Practice

Rosanna Salvia[1], Alessia D'Agata[2,3], Adele Sateriano[4], Marco Maialetti[5], Luca Salvati[2], Ioannis Konaxis[6], and Giovanni Quaranta[1]

[1]Department of Mathematics, Informatics and Economics, University of Basilicata, Italy
[2]Department of Methods and Models for Economics, Territory and Finance (MEMOTEF), Faculty of Economics, Sapienza University of Rome, Italy
[3]Department of Geography, Building B, Autonomous University of Barcelona, Spain
[4]Mediterranean Foundation for Sustainable Development (MEDES), Italy
[5]Independent Researcher and Lawyer, Italy
[6]Department of Tourism Studies, University of Piracus, Greece

Email: rosanna.salvia@unibas.it; alessia.dagata@uniroma1.it; adele@medes.eu; avv.marcomaialetti@gmail.com; luca.salvati@uniroma1.it; ikonaxis@unipi.gr; giovanni.quaranta@unibas.it

Abstract

Agro-environmental degradation has emerged as a critical global issue, particularly given agriculture's dual role as both a driver and victim of environmental change. Unsustainable farming practices, land-use transformations, and socioeconomic pressures have accelerated soil degradation, leading to compaction, erosion, nutrient depletion, and biodiversity loss. Since the mid-20th century, processes of agricultural specialization, urban expansion, and rural abandonment have reshaped landscapes, exacerbating conflicts over natural resources and intensifying pressures on soil and water systems. Key manifestations of degradation include pollution from pesticides and fertilizers, desertification, and the homogenization of agro-ecosystems through industrial crops. Moreover, climate change – closely linked to land-use

DOI: 10.1201/9788743811107-2

practices – has amplified these threats, contributing to a sudden increase in air and soil temperatures, altered hydrological cycles, and higher risks of extreme events, with profound implications for Mediterranean – and, more generally, European – regions. Soil degradation and desertification, recognized as major environmental priorities, further threaten ecosystem services, food security, and human well-being. Habitat fragmentation and land consumption compound these dynamics by accelerating biodiversity loss and reducing ecological resilience. Overall, the evidence underscores the urgent need for integrated strategies that address the interplay between agriculture, land management, and climate change in order to mitigate agro-environmental degradation and safeguard socioecological systems.

2.1 Introduction

Agro-environmental degradation is a relevant issue, as agriculture is the productive sector most affected by recent environmental problems (Adams et al., 2004). This impact can be understood in two opposing ways: on the one hand, agricultural practices contribute to protecting the soil from degradation processes; on the other hand, where unsustainable agricultural models are adopted, soil degradation processes are accelerated (Bajocco et al., 2012; Bajocco et al., 2015; Biasi et al., 2015). Agricultural production systems have been affected by progressive specialization trends, as well as the concomitant marginalization of areas morphologically and geographically suboptimal for intensive crop production (Ramsfield et al., 2016). This has triggered processes of social exclusion and the resulting differentiation between agricultural areas exposed to land abandonment and highly specialized and productive areas (Salvati & Carlucci, 2015).

The expansion of industrial practices and the resulting homogenization of the agricultural landscapes represent a possible response to the specific economic dynamics that have driven production toward intense agricultural intensification (Grossman & Krueger, 1995; Arrow et al., 1996; Dinda, 2004). Mechanical tillage practices – repeated irrationally over decades – negatively impact soil structure and fertility, leading to degradation phenomena such as compaction, erosion, nitrate leaching, and organic matter depletion (Cerdà et al., 2021). Agricultural land, like natural environments, has suffered the negative impact of settlement processes over the last decades (Turner et al., 2003; Salvati, 2013, 2014). As a result of intense economic growth and social development since the 1950s, arable land has shrunk by approximately a quarter (Dietz & Rosa, 1994).

On the one hand, there has been a consolidation of small plots (possibly no longer competitive on the market) and, on the other hand, a transformation from agricultural to residential uses has been progressively observed (Salvati & Ranalli, 2015). In terms of land use, the rural exodus was more intensively associated with the concentration of residential/industrial settlements and agricultural activities in lowlands, and the marginalization of hilly and mountainous areas (e.g., Treweek, 1996). Urban and industrial concentration in formerly agricultural areas creates conflicts for land use and primary resources, namely soil, water, and clean air (Fisher & Christopher, 2007).

Agricultural soils have therefore been subjected to a series of degradation processes and threats, such as erosion, reduction in organic matter, pollution, soil sealing, compaction, loss of biodiversity, salinization, flooding, and landslides (Chapin III et al., 2000; Chase et al., 2020; Jaureguiberry et al., 2022). Pollution caused by pesticides and synthetic fertilizers used in agricultural practices is one of the most widespread forms of agro-environmental degradation (Rosa & Dietz, 1998). Additionally, the introduction of chemicals into the soil has negative effects on water and air quality, thus posing a threat to biodiversity and human health (Roy & Tomar, 2000; Raymond et al., 2009; Rodrigo-Comino et al., 2021). Chemical substances that constitute pesticides and that are used to chemically control organisms that damage agricultural systems first emerged in the 1950s (Randin et al., 2020). Controlling organisms that could conflict with agricultural production (such as fungi, plants, and insects) with pesticides contributes to ecosystem and socioeconomic damage (human health).

2.2 Climate Change

More than 20 years ago, the first world conference on climate change was held in Geneva with the aim of predicting and preventing potential climate changes caused by human activities that could have a negative impact on human well-being (Rosa & Dietz, 1998). The issue that is causing the greatest concern is the unprecedented rising trend in global average temperatures (up 0.6 °C in the last 100 years) and the assumption that regional warming is not solely due to natural causes (Rosa et al., 2004).

Over the last century, humans have caused profound changes in the composition of the earth's atmosphere, resulting in an unprecedented increase in the concentrations of certain chemical substances (Dietz & Rosa, 1994; Dasgupta et al., 2002; York et al., 2003). Even small variations in the concentrations of these components can alter the balance of landscape systems, both globally and regionally (Fischer-Kowalski & Amann, 2001). The increase in the greenhouse effect in recent years has caused (and will cause) a series of physical changes in the atmosphere that could lead to an increase in average temperatures, and consequently, the melting of glaciers and ice caps (Chapin III et al., 2000).

According to the Intergovernmental Panel on Climate Change (IPCC), the increase in greenhouse gas concentrations over the last century is attributable to emissions from the burning of fossil fuels such as coal, oil, and natural gas (Grossman & Krueger, 1995; Arrow et al., 1996; Dinda, 2004). Furthermore, the IPCC states that climate change has already had its effects. Further impacts may occur in the coming years on human health, food security, socioeconomic development, water resources, and infrastructure, if immediate response actions are not taken (Fisher & Christopher, 2007).

In 1992, during the United Nations Conference on Environment and Development, the United Nations Framework Convention on Climate Change (UNFCCC) was opened for signature (Giddings et al., 2002). The signatory governments committed, albeit in a non-binding manner, to reducing atmospheric emissions in order to prevent interference between the human and climate systems (Ehrlich & Holdren, 1971). The first Conference of the Parties (COP1) was held in Berlin in 1995, where it was decided to identify actions consistent with the objectives of the UNFCCC (Holdren & Ehrlich, 1974): reforestation, for instance, was understood as a viable action to store atmospheric carbon and offset greenhouse gas emissions (Mather, 1992). The Marrakesh Conference (COP22, held in November 2016) defined the implementation aspects of the Paris Agreement, involving the signatory countries of the Kyoto Protocol, which entered into force in 2005 (Arrow et al., 1996; Grossman & Krueger, 1995; Adams et al., 2004).

Globally, it has been noted that changes in land use and management are currently contributing to approximately 25% of global anthropogenic carbon emissions (Ramsfield et al., 2016). The terrestrial biosphere plays a fundamental role in the carbon cycle: it exchanges large quantities with the atmosphere via plants, which capture large amounts of CO_2 from the atmosphere through photosynthesis and release a portion through respiration (Turner et al., 2003). Additional CO_2 is also fixed in the living matter of plants and animals, as well as in litter and soil (Roy & Tomar, 2000).

The increase in carbon dioxide concentration in the atmosphere over the last hundred years reveals how human activities are impacting climate change and the earth's ability to restore its equilibrium (Rodrigo-Comino et al., 2021). The alteration of CO_2 concentration in the atmosphere is caused by various factors, such as the removal of forest areas in favor of intensive agriculture and livestock farming (e.g., the Amazon rainforest), which in turn leads to a reduction in the carbon absorption capacity of terrestrial ecosystems and an increase in atmospheric emissions (Randin et al., 2020). The knock-on effects of climate change and the greenhouse effect include, for instance, an increase in average temperatures, rising sea levels, the melting of glaciers and polar ice caps, and the increased frequency of devastating weather events (Rosa & Dietz, 1998).

Global warming projected by climate scenarios reflects increasing average annual temperatures, reducing summer precipitation, a general increase in average temperatures, and increased dry spells, with serious consequences for local economies, agricultural traditions, and identities (Rosa et al., 2004; Bajocco et al., 2012; Bajocco et al., 2015). This will also expose ecosystems to substantial changes in water availability (Jaureguiberry et al., 2022). In this perspective, environmental degradation has become a real threat precisely because of the impact of anthropogenic changes (Salvati, 2023).

A series of variables, detailed below, has raised recent concerns. Land use transformations, from natural (forests and wetlands) to semi-natural (cultivated) or artificial (construction, industry, infrastructure) uses, are among the most obvious (Salvati & Carlucci, 2014). These changes, in addition to causing the often permanent and irreversible loss of fertile soil, determine further negative impacts, such as land fragmentation, reduced biodiversity, alterations in hydrogeological cycles, and various microclimatic changes (Chase et al., 2020).

The spread of settlements, along with the associated increase in land infrastructure, is driving increased transportation needs and energy consumption, contributing to the worrying increase in average global temperatures and greenhouse gas emissions (Esposito et al., 2016; Smiraglia et al., 2016; Sallustio et al., 2018). This will expose ecosystems to substantial changes in water availability (Johnson et al., 2017). Climate change definitely impacts the stability of ecosystems, causing imbalances and instability, and resulting in a loss of biodiversity (Hanski, 2011). Changes in the distribution of flora and fauna are currently the focus of attention in the scientific community, with regard to the impact on humans and health (Jaureguiberry et al., 2022).

2.3 Soil Degradation

Desertification is a process that leads to an irreversible reduction in the soil's capacity to produce resources and services that ensure the subsistence of biomass (Esposito et al., 2016). An earlier definition, based on a criterion of biological productivity, identifies this phenomenon as a "reduction (or destruction) of the soil's biological potential that can lead to desert conditions" (United Nations Conference on Desertification, 1977) and is independent of the geographical location of the affected areas, their climatic characteristics, the causes (natural or anthropogenic), and the processes (salinization, erosion, deforestation) that cause the degradation of the soil's biological potential (Biasi et al., 2015; Kelly et al., 2015; Halbac-Cotoara-Zamfir et al., 2020). In 1984, desertification was defined by the UNCCD (United Nations Convention to Combat Desertification) as

> *"the irreversible process of land degradation in arid, semi-arid and dry sub-humid zones, caused by various factors, including climatic variations and the relationship with the current pattern of human activities; it involves a decrease or disappearance of the productivity and biological or economic complexity of cultivated lands (irrigated or not), grasslands, pastures, forests or woodlands as a result of land use or of one or more phenomena related to human activity and settlement patterns" (Article 1.a).*

The concept of "desertification" has gradually evolved over the years in an attempt to define a process that, although triggered by local causes, is increasingly becoming a global problem (Bruno et al., 2024). A common element that unequivocally associates areas subject to desertification is the progressive reduction of the topsoil layer and its productive capacity (Quaranta et al., 2023). The UNCCD has therefore chosen to adopt a definition that limits its scope of territorial intervention based on climatic characteristics and explicitly includes climate change as one of the causes of the phenomenon, in addition to human action (Kalogeropoulos et al., 2024).

Desertification is therefore defined as a form of land degradation, basically observed in point and areal locations (namely, arid, semi-arid, and dry sub-humid districts), and triggered by climate warming and rising human pressure (Egidi et al., 2021). It begins in localized areas and progresses in patches and in successive stages, undergoing abrupt deteriorations during particularly dry periods or regressions during wetter ones (Imbrenda et al., 2022; Imbrenda et al., 2023). This phenomenon should not be considered only in its final stage, but rather as a complex process triggered and fueled by the combination of several phenomena, such as (i) soil erosion, (ii) change in soil structural parameters, (iii) salinization, (iv) removal of vegetation cover and regenerative material, (v) changes in rainfall patterns, and (vi) interactions between the land surface and the atmosphere (Ferrara et al., 2020; Cerdà et al., 2021; Baartman et al., 2022).

Land degradation, therefore, means a reduction in biological and/or economic productivity and the complexity of crops, pastures, and forests, due to land use (e.g., irrational agricultural practices, intensive grazing, deforestation, wildfires, and urbanization) (Kairis et al., 2015; Karamesouti et al., 2015; Kosmas et al., 2015). The processes that affect soils are essentially alterations to the physical, chemical, and biological properties or the destruction of vegetation cover (Canfora et al., 2017). While not necessarily implying the expansion of deserts, desertification leads to a drastic reduction in soil fertility and, consequently, in an ecosystem's capacity to provide services (Salvati et al., 2015). The term "desertification" therefore refers to a generic degradation of land in specific climate zones (Mosconi et al., 2020).

The factors and causes influencing this complex phenomenon are multiple and diverse (Biasi et al., 2017): while it is true that underlying natural factors predispose a given area to degradation processes, it is also true that environmental conditions and human intervention play a significant role as driving forces behind this process (D'Agata et al., 2024). Fragile ecosystems, clayey-sandy sedimentary soils, shallow aquifers with little or no vegetation cover, and already compromised areas are environments particularly susceptible to erosion and degradation, where even a small alteration in ecological conditions can trigger irreversible desertification processes (Brunori et al., 2017).

Such changes can arise from the presence of particular climatic conditions. Rising temperatures are considered one of the natural drivers of desertification because, by increasing the rate of evapotranspiration and thus reducing the soil's water-retaining capacity, they often cause a drying out of local soil and climate conditions – especially when accompanied by a significant decrease in precipitation (Lin et al., 2017; Usman & Hammar, 2021; Baartman et al., 2022). Both aridity and drought contribute to making a territory less suitable to living organisms, and the resulting absence or sparse presence of plant species covering the soil in turn makes the land more exposed to wind and water erosion (Pili et al., 2019; Mancini et al., 2018). In such scenarios, intense rainfall events paradoxically constitute another risk of early desertification (Samela et al., 2022). Soils, no longer able to absorb water efficiently because they are parched and lack adequate vegetation cover, are subject to intense erosion and leaching of the most fertile surface layers rich in organic matter, a process that would increase landscape degradation (Bajocco et al., 2012).

Human intervention plays a key role in triggering degradation processes (Bajocco et al., 2015; Quaranta et al., 2020, 2023). Wildfires, intensive agricultural practices, unsustainable livestock grazing, and land take all contribute to the chemical and physical deterioration of fertile soils (Fares et al., 2017). Worldwide, 24 billion tons of cultivated land are lost each year due to desertification (Bruno et al., 2024). In Europe, over 20 million hectares of land have been degraded due to industrial wastewater and acid rain, and as much as 25% of agricultural land and 35% of pastureland are currently at risk (Kalogeropoulos et al., 2024). The Northern Mediterranean basin is undoubtedly the European region most exposed to this problem; here, intense water exploitation, agricultural abandonment, and excessive human activity are accompanied by prolonged spells of drought, high soil erodibility, and a high frequency (and severity) of wildfires (Mancino et al., 2016).

2.3.1 Early desertification processes

Regarding the quantitative study of desertification, the United Nations Convention to Combat Desertification (UNCCD) represents the primary reference for the global scientific community, establishing applied research strategies at the continental, national, and local levels and assessing desertification risk through interpretative models (Ferrara et al., 2020; Halbac-Cotoara-Zamfir et al., 2020; Imbrenda et al., 2022). The most widely applied model, which has become a reference standard in the literature because it is designed to identify desertification risk through composite indicators, is the Environmentally Sensitive Areas (ESA) model (Esposito et al., 2016; Francaviglia et al., 2019; Salvia et al., 2023). This logical approach defines a concept of environmental quality reflected in different components: soil, climate, vegetation, and land management (Baartman et al., 2022). For each component, a certain number of factors deemed significant in explaining desertification processes are identified, assigning them a specific quality class and reference score (Karamesouti et al., 2015; D'Agata et al., 2024; Bruno et al., 2024).

For each of these components, a thematic quality index is developed that is a function of the factors that characterize it (Kairis et al., 2015). Finally, this set of indicators is composed into an ESA index, which expresses the level of vulnerability to desertification (Kelly et al., 2015; Kosmas et al., 2015). However, there are other important conceptual models for the study of desertification (Figure 2.1), including the DPSIR framework ("Driving forces, Pressures, State, Impacts and Responses," developed in 1994 by the Organization for Economic Co-operation and Development, OECD). It identifies the causal linkage between anthropogenic pressures, environmental impacts, and responses aimed at mitigating these impacts (Imbrenda et al., 2023). In both models, the aridity index represents a key element in defining the climatic aspect of the phenomenon (Cerdà et al., 2021).

Desertification is now recognized as one of the most serious environmental priorities affecting arid, semi-arid, and sub-humid lands (Brunori et al.,

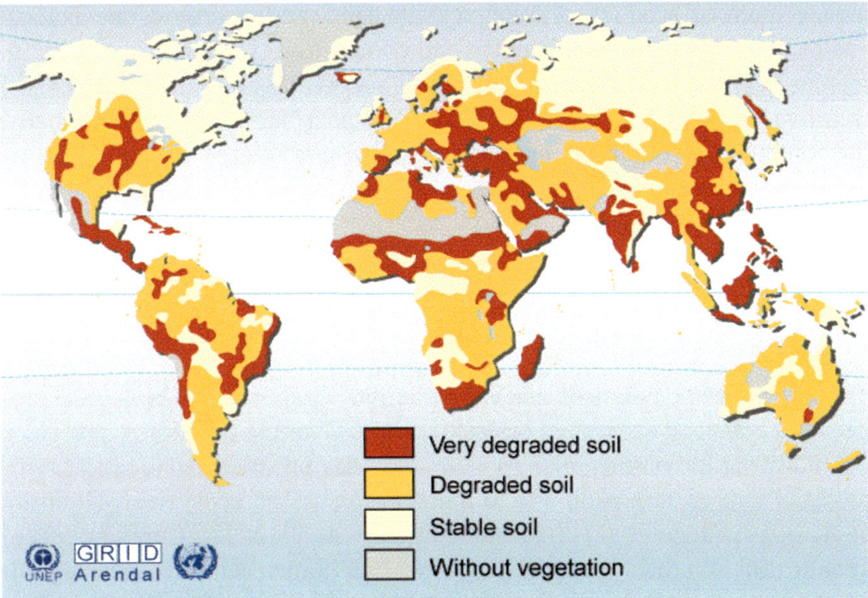

Figure 2.1 A global map of soil degradation. Source: authors' elaboration of the UNEP/GRID dataset.

2017). Although the desertification process in advanced economies seems to be very slow and gradual, even in its intermediate stages (before the establishment of a true desert), significant changes in soil characteristics can be observed, with soil fertility drastically decreasing, sometimes to the point of exceeding the minimum levels required for agricultural use (Baartman et al., 2022). The causes of this intensification can be attributed both to anthropogenic activities (increasingly widespread and invasive) and to factors inherent in the very nature of the environment in which they occur (Biasi et al., 2015, 2017; Fares et al., 2017). Excessively dry and hot summers (which dry out the soil, making it dry and impermeable) followed by periods of short, intense rainfall, can cause real geological and ecological disasters, leading to soil erosion and leaching that result in a loss of fertility and productivity (Mancino et al., 2016; Marchi et al., 2018; Egidi et al., 2021).

Overexploitation of water resources is compromising the abundance of groundwater, also leading to the salinization of underground aquifers along the coasts (Canfora et al., 2017). Wildfires negatively impact the composition and structure of plant and animal communities and the physical and chemical properties of the soil, changing its structure and making it less permeable, and thus, more susceptible to erosion (Mancini et al., 2018). Incorrect management of production resources and land, and the use of mechanical means, lead to soil impoverishment and compaction (Imbrenda et al., 2022). Increasingly intensive farming causes environmental pollution due to excessive livestock density in limited areas (Karamesouti et al., 2015). Finally, the continued expansion of urbanized areas contributes to desertification by removing fertile land from agricultural use, harming soil production capacity (Masini et al., 2019).

2.3.2 Biodiversity, landscape fragmentation, and soil consumption

The natural phenomenon of species extinction is a slow-paced biological process in a balanced ecosystem (Hanski, 2011). Over the past 150 years, since the Industrial Revolution, many species have disappeared, and others are currently at risk of extinction due to human pressure on ecosystems (Johnson et al., 2017; Chase et al., 2020; Jaureguiberry et al., 2022). Current concern about extinction rates stems from the fact that human activities have significantly accelerated this natural process (Chapin III et al., 2000). Human action has radically (and often irreversibly) altered the structure and functionality of natural ecosystems, compromising the wellbeing of resident populations (Blackmore & Hawksworth, 1995). The greatest threats to biological diversity arise from the impact of human activities on habitats and biocoenoses (Arrow et al., 1996; Adams et al., 2004; Fisher & Christopher, 2007). In many cases, environmental damage results directly from industrial and commercial activities associated with the global economy, such as deforestation for intensive agriculture or urban development, wetland reclamation, dam construction, and highway construction (Roy & Tomar, 2000).

The disappearance of natural environments can occur through outright destruction or slower soil deterioration due to land parcel fragmentation and industrialization of agriculture (Turner et al., 2003). The latter is a process by which a natural or semi-natural habitat, which continuously occupies a large area, decreases in size and is divided into small portions or fragments (Randin et al., 2020). This almost always occurs in conjunction with interventions that drastically reduce the area occupied by a given habitat, but can also be caused by parallel events that have a strong impact on the environment (Assennato et al., 2022): roads, railways, airports, irrigation canals, power lines, pipelines, cultivated fields, fences, or other types of barriers that limit or completely prevent the movement of individuals, dividing populations and reducing gene flow between them (Dietz & Rosa, 1994; Rosa & Dietz, 1998; Rosa et al., 2004).

From an ecological perspective, fragmented habitats are significantly more susceptible to external threats than non-fragmented habitats (Dutcher et al., 2007). The former, displaying a longer perimeter and therefore a larger marginal habitat, has an area more exposed to neighboring areas altered (or degraded) by humans (Giddings et al., 2002). Fragmentation brings with it multiple consequences. Indeed, in addition to reducing dispersal, food availability can decrease, for instance, forcing species to continually rely on the same area, leading to a reduction in natural resources and the irreversible degradation of habitats and its communities. This series of consequences resulting from fragmentation processes increases the vulnerability of fragmented habitats (Chapin III et al., 2000; Turner et al., 2003; Raymond et al., 2009).

The intrinsic fragmentation of natural environments is considered one of the main anthropogenic threats to biodiversity (Blackmore & Hawksworth, 1995). The destruction and transformation of natural environments, their reduction in size, value, and functions, and increased isolation, all affect the structure and dynamics of sensitive (animal and plant) populations, up to the point of altering community parameters, systemic functions, and ecological processes (Jaureguiberry et al., 2022). In addition to the actual loss of habitats, a consequence of human settlements and infrastructure construction (e.g., roads, railways, harbors, and airports) is the creation of barriers to species dispersal (Randin et al., 2020). Changes in density, temperature, atmospheric gas pollution, erosion, sedimentation, and waste accumulation are just a few examples of changes in the physical environment that can alter the distribution of biological species (Rosa et al., 2004).

However, these barriers often serve a dual purpose: while they partially or completely interrupt the flow of some species, they can also create biological corridors for other marginal (but generalist) species (Chase et al., 2020). Thematic maps, satellite imagery, and Geographic Information Systems, are useful tools for initial assessments of the type and degree of fragmentation (Roy & Tomar, 2000). Appropriately processed spatial data can identify contiguities (or discontinuities) between landscape elements and serve as a fundamental starting point for subsequent analyses.

2.4 Concluding Remarks

As a dynamic system, soil is subject to continuous exchanges of matter and energy with surrounding systems. However, anthropogenic influence on this limited resource highlights the need for quantitative tools, such as indicators, to study how humans interact with the surface layer of the earth's crust, on which it depends. Human activities, in fact, influence soil functionality and its ability to support agricultural productivity. Intensive agriculture, by introducing significant alterations in nutrient cycles and soil regulation processes, is one of the major causes of agro-environmental degradation. This degradation, in turn, has significant effects on global climate change, desertification, biodiversity loss, and landscape fragmentation.

References

Adams, W. M., Aveling, R., Brockington, D., Dickson, B., Elliott, J., Hutton, J., Roe, D., Vira, B., & Wolmer, W. (2004). Biodiversity Conservation and the Eradication of Poverty. Science, 306(5699), 1146–1149.

Arrow, K., Bolin, B., Costanza, R., Dasgupta, P., Folke, C., Holling, C. S., Jansson, B.-O., Levin, S., Mäler, K.-G., Perrings, C., & Pimentel, D. (1996). Economic growth, carrying capacity, and the environment. Environment and Development Economics, 1(1), 104–110.

Assennato, F., Smiraglia, D., Cavalli, A., Congedo, L., Giuliani, C., Riitano, N., Strollo, A., & Munafò, M. (2022). The Impact of Urbanization on Land: A Biophysical-Based Assessment of Ecosystem Services Loss Supported by Remote Sensed Indicators. Land, 11(2), 236.

Baartman, J. E. M., Nunes, J. P., van Delden, H., Vanhout, R., & Fleskens, L. (2022). The Effects of Soil Improving Cropping Systems (SICS) on Soil Erosion and Soil Organic Carbon Stocks across Europe: A Simulation Study. Land, 11(6), 943.

Bajocco, S., De Angelis, A., & Salvati, L. (2012). A satellite-based green index as a proxy for vegetation cover quality in a Mediterranean region. Ecological Indicators, 23, 578–587.

Bajocco, S., Dragoz, E., Gitas, I., Smiraglia, D., Salvati, L., & Ricotta, C. (2015). Mapping Forest Fuels through Vegetation Phenology: The Role of Coarse-Resolution Satellite Time-Series. PLOS ONE, 10(3), e0119811.

Biasi, R., Brunori, E., Ferrara, C., & Salvati, L. (2017). Towards sustainable rural landscapes? a multivariate analysis of the structure of traditional tree cropping systems along a human pressure gradient in a mediterranean region. Agroforestry Systems, 91(6), 1199–1217.

Biasi, R., Brunori, E., Smiraglia, D., & Salvati, L. (2015). Linking traditional tree-crop landscapes and agro-biodiversity in central Italy using a database of typical and traditional products: a multiple risk assessment through a data mining analysis. Biodiversity and Conservation, 24(12), 3009–3031.

Blackmore, S., & Hawksworth, D. L. (1995). Biodiversity: Measurement and Estimation. The Journal of Ecology, 83(5), 911.

Bruno, E., Salvia, R., Quaranta, G., Cudlin, P., Punzo, G., & Salvati, L. (2024). Identifying On-Site and Off-Site Drivers of Land Degradation in Advanced Economies: A Spatial Approach for Italy. Environmental and Resource Economics, 87(9), 2429–2453.

Brunori, E., Salvati, L., Mancinelli, R., Smiraglia, D., & Biasi, R. (2017). Multi-temporal land use and cover changing analysis: the environmental impact in Mediterranean area. International Journal of Sustainable Development & World Ecology, 24(3), 276–288.

Burnett, C., & Blaschke, T. (2003). A multi-scale segmentation/object relationship modelling methodology for landscape analysis. Ecological Modelling, 168(3), 233–249.

Canfora, L., Salvati, L., Benedetti, A., & Francaviglia, R. (2017). Is soil microbial diversity affected by soil and groundwater salinity? Evidences from a coastal system in central Italy. Environmental Monitoring and Assessment, 189(7), 319.

Cerdà, A., Daliakopoulos, I. N., Terol, E., Novara, A., Fatahi, Y., Moradi, E., Salvati, L., & Pulido, M. (2021). Long-term monitoring of soil bulk density and erosion rates in two Prunus Persica (L) plantations under flood irrigation and glyphosate herbicide treatment in La Ribera district, Spain. Journal of Environmental Management, 282, 111965.

Chapin III, F. S., Zavaleta, E. S., Eviner, V. T., Naylor, R. L., Vitousek, P. M., Reynolds, H. L., Hooper, D. U., Lavorel, S., Sala, O. E., Hobbie, S. E., Mack, M. C., & Díaz, S. (2000). Consequences of changing biodiversity. Nature, 405(6783), 234–242.

Chase, J. M., Blowes, S. A., Knight, T. M., Gerstner, K., & May, F. (2020). Ecosystem decay exacerbates biodiversity loss with habitat loss. Nature, 584(7820), 238–243.

Cheminal, A., Kokkoris, I. P., Zotos, A., Strid, A., & Dimopoulos, P. (2022). Assessing the Ecosystem Services Potential of Endemic Floras: A Systematic Review on the Greek Endemics of Peloponnese. Sustainability, 14(10), 5926.

D'Agata, A., Cudlin, P., Vardopoulos, I., Schinaia, G., Corona, P., & Salvati, L. (2024). Assessing the spatial coherence of forest cover indicators from different data sources: A contribution to sustainable development reporting. Ecological Indicators, 158, 111498.

Dasgupta, S., Laplante, B., Wang, H., & Wheeler, D. (2002). Confronting the Environmental Kuznets Curve. Journal of Economic Perspectives, 16(1), 147–168.

Dietz, T., & Rosa, E. A. (1994). Rethinking the Environmental Impacts of Population, Affluence and Technology. Human Ecology Review, 1(2), 277–300.

Dinda, S. (2004). Environmental Kuznets Curve Hypothesis: A Survey. Ecological Economics, 49(4), 431–455.

Doukas, Y. E., Salvati, L., & Vardopoulos, I. (2023). Unraveling the European agricultural policy sustainable development trajectory. Land, 12(9), 1749.

Dutcher, D. D., Finley, J. C., Luloff, A. E., & Johnson, J. B. (2007). Connectivity With Nature as a Measure of Environmental Values. Environment and Behavior, 39(4), 474–493.

Egidi, G., Cividino, S., Paris, E., Palma, A., Salvati, L., & Cudlin, P. (2021). Assessing the impact of multiple drivers of land sensitivity to desertification in a Mediterranean country. Environmental Impact Assessment Review, 89, 106594.

Ehrhardt-Martinez, K., Crenshaw, E. M., & Jenkins, J. C. (2002). Deforestation and the Environmental Kuznets Curve: A Cross-National Investigation of Intervening Mechanisms. Social Science Quarterly, 83(1), 226–243.

Ehrlich, P. R., & Holdren, J. P. (1971). Impact of Population Growth. Science, 171(3977), 1212–1217.

Esposito, P., Patriarca, F., Perini, L., & Salvati, L. (2016). Land degradation, economic growth and structural change: evidences from Italy. Environment, Development and Sustainability, 18(2), 431–448.

Fabbio, G., Cantiani, P., Ferretti, F., di Salvatore, U., Bertini, G., Becagli, C., Chiavetta, U., Marchi, M., & Salvati, L. (2018). Sustainable Land Management, Adaptive Silviculture, and New Forest Challenges: Evidence from a Latitudinal Gradient in Italy. Sustainability, 10(7), 2520.

Fares, S., Bajocco, S., Salvati, L., Camarretta, N., Dupuy, J.-L., Xanthopoulos, G., Guijarro, M., Madrigal, J., Hernando, C., & Corona, P. (2017). Characterizing potential wildland fire fuel in live vegetation in the Mediterranean region. Annals of Forest Science, 74(1), 1.

Ferrara, A., Kosmas, C., Salvati, L., Padula, A., Mancino, G., & Nolè, A. (2020). Updating the MEDALUS ESA Framework for Worldwide Land Degradation and Desertification Assessment. Land Degradation & Development, 31(12), 1593–1607.

Fischer-Kowalski, M., & Amann, C. (2001). Beyond IPAT and Kuznets Curves: Globalization as a Vital Factor in Analysing the Environmental Impact of Socio-Economic Metabolism. Population and Environment, 23(1), 7–47.

Fisher, B., & Christopher, T. (2007). Poverty and biodiversity: Measuring the overlap of human poverty and the biodiversity hotspots. Ecological Economics, 62(1), 93–101.

Forino, G., Ciccarelli, S., Bonamici, S., Perini, L., & Salvati, L. (2015). Developmental Policies, Long-Term Land-Use Changes and the Way Towards Soil Degradation: Evidence from Southern Italy. Scottish Geographical Journal, 131(2), 123–140.

Francaviglia, R., Di Bene, C., Farina, R., Salvati, L., & Vicente-Vicente, J. L. (2019). Assessing "4 per 1000" soil organic carbon storage rates under Mediterranean climate: a comprehensive data analysis. Mitigation and Adaptation Strategies for Global Change, 24(5), 795–818.

Freeman III, A. M., Herriges, J. A., & Kling, C. L. (2014). The Measurement of Environmental and Resource Values. Routledge, London.

Giddings, B., Hopwood, B., & O'Brien, G. (2002). Environment, economy and society: fitting them together into sustainable development. Sustainable Development, 10(4), 187–196.

Golia, E. E., & Zorpas, A. (2022). Assessing environmental impacts through innovative solutions. Environmental Science and Pollution Research, 29(20), 29524–29527.

Grossman, G. M., & Krueger, A. B. (1995). Economic Growth and the Environment. The Quarterly Journal of Economics, 110(2), 353–377.

Halbac-Cotoara-Zamfir, R., Polinesi, G., Chelli, F., Salvati, L., Bianchini, L., Marucci, A., & Colantoni, A. (2022). Found in Complexity, Lost in Fragmentation: Putting Soil Degradation in a Landscape Ecology Perspective. International Journal of Environmental Research and Public Health, 19(5), 2710.

Halbac-Cotoara-Zamfir, R., Smiraglia, D., Quaranta, G., Salvia, R., Salvati, L., & Giménez-Morera, A. (2020). Land Degradation and Mitigation Policies in the Mediterranean Region: A Brief Commentary. Sustainability, 12(20), 8313.

Hanski, I. (2011). Habitat Loss, the Dynamics of Biodiversity, and a Perspective on Conservation. Ambio, 40(3), 248–255.

Holdren, J. P., & Ehrlich, P. R. (1974). Human Population and the Global Environment: Population growth, rising per capita material consumption, and disruptive technologies have made civilization a global ecological force. American Scientist, 62(3), 282–292.

Imbrenda, V., Coluzzi, R., Di Stefano, V., Egidi, G., Salvati, L., Samela, C., Simoniello, T., & Lanfredi, M. (2022). Modeling Spatio-Temporal Divergence in Land Vulnerability to Desertification with Local Regressions. Sustainability, 14(17), 10906.

Imbrenda, V., Coluzzi, R., Mariani, F., Nosova, B., Cudlinova, E., Salvia, R., Quaranta, G., Salvati, L., & Lanfredi, M. (2023). Working in (Slow) Progress: Socio-Environmental and Economic Dynamics in the Forestry Sector and the Contribution to Sustainable Development in Europe. Sustainability, 15(13), 10271.

Jakubínský, J., Prokopová, M., Raška, P., Salvati, L., Bezak, N., Cudlín, O., Cudlín, P., Purkyt, J., Vezza, P., Camporeale, C., Daněk, J., Pástor, M., & Lepeška, T. (2021). Managing floodplains using nature-based solutions to support multiple ecosystem functions and services. WIREs Water, 8(5).

Jaureguiberry, P., Titeux, N., Wiemers, M., Bowler, D. E., Coscieme, L., Golden, A. S., Guerra, C. A., Jacob, U., Takahashi, Y., Settele, J., Díaz, S., Molnár, Z., & Purvis, A. (2022). The direct drivers of recent global anthropogenic biodiversity loss. Science Advances, 8(45).

Johnson, C. N., Balmford, A., Brook, B. W., Buettel, J. C., Galetti, M., Guangchun, L., & Wilmshurst, J. M. (2017). Biodiversity losses and conservation responses in the Anthropocene. Science, 356(6335), 270–275.

Kairis, O., Karavitis, C., Salvati, L., Kounalaki, A., & Kosmas, K. (2015). Exploring the Impact of Overgrazing on Soil Erosion and Land Degradation in a Dry Mediterranean Agro-Forest Landscape (Crete, Greece). Arid Land Research and Management, 29(3), 360–374.

Kalogeropoulos, K., Tsesmelis, D., Tsatsaris, A., Zervas, E., Karavitis, C., Vasilakou, C., & Barouchas, P. (2024). Geospatial insights into Greece's desertification vulnerability: a composite indicator approach. GeoHazards, 5(2), 374–392.

Karamesouti, M., Detsis, V., Kounalaki, A., Vasiliou, P., Salvati, L., & Kosmas, C. (2015). Land-use and land degradation processes affecting soil resources: Evidence from a traditional Mediterranean cropland (Greece). Catena, 132(45–55).

Kelly, C., Ferrara, A., Wilson, G. A., Ripullone, F., Nolè, A., Harmer, N., & Salvati, L. (2015). Community resilience and land degradation in forest and shrubland socio-ecological systems: Evidence from Gorgoglione, Basilicata, Italy. Land Use Policy, 46, 11–20.

Kosmas, C., Detsis, V., Karamesouti, M., Kounalaki, K., Vassiliou, P., & Salvati, L. (2015). Exploring Long-Term Impact of Grazing Management on Land Degradation in the Socio-Ecological System of Asteroussia Mountains, Greece. Land, 4(3), 541–559.

Lin, S., Wang, S., Marinova, D., Zhao, D., & Hong, J. (2017). Impacts of urbanization and real economic development on CO2 emissions in non-high income countries: Empirical research based on the extended STIRPAT model. Journal of Cleaner Production, 166, 952–966.

Mancini, L. D., Corona, P., & Salvati, L. (2018). Ranking the importance of Wildfires' human drivers through a multi-model regression approach. Environmental Impact Assessment Review, 72, 177–186.

Mancino, G., Nolè, A., Salvati, L., & Ferrara, A. (2016). In-between forest expansion and cropland decline: A revised USLE model for soil erosion risk under land-use change in a Mediterranean region. Ecological Indicators, 71, 544–550.

Marchi, M., Ferrara, C., Biasi, R., Salvia, R., & Salvati, L. (2018). Agro-Forest Management and Soil Degradation in Mediterranean Environments: Towards a Strategy for Sustainable Land Use in Vineyard and Olive Cropland. Sustainability, 10(7), 2565.

Masini, E., Tomao, A., Barbati, A., Corona, P., Serra, P., & Salvati, L. (2019). Urban growth, land-use efficiency and local socioeconomic context: a comparative analysis of 417 Metropolitan regions in Europe. Environmental Management, 63(3), 322–337.

Mather, A. S. (1992). The Forest Transition. Area, 24(4), 367–379.

Morgan, R. K. (2012). Environmental impact assessment: the state of the art. Impact Assessment and Project Appraisal, 30(1), 5–14.

Morrison-Saunders, A. (1999). Exploring the EIA/Environmental Management Relationship. Environmental Management, 24(3), 281–295.

Morrow, N., Salvati, L., Colantoni, A., & Mock, N. (2018). Rooting the future; On-farm trees' contribution to household energy security and asset creation as a resilient development pathway—evidence from a 20-year panel in rural Ethiopia. Sustainability, 10(12), 4716.

Mosconi, E. M., Colantoni, A., Gambella, F., Cudlinová, E., Salvati, L., & Rodrigo-Comino, J. (2020). Revisiting the Environmental Kuznets Curve: The Spatial Interaction between Economy and Territory. Economies, 8(3), 74.

Nickayin, S. S., Quaranta, G., Salvia, R., Cividino, S., Cudlin, P., & Salvati, L. (2021). Reporting land degradation sensitivity through multiple indicators: Does scale matter? Ecological Indicators, 125, 107560.

Nita, A. (2019). Empowering impact assessments knowledge and international research collaboration - A bibliometric analysis of Environmental Impact Assessment Review journal. Environmental Impact Assessment Review, 78, 106283.

Panagiotopoulos, P., Vardopoulos, I., Maialetti, M., Ciaschini, C., Koundouri, P., & Salvati, L. (2024). Reimagining sustainable development and economic performance indicators: a human-centric Maslow–Bossel blueprint. Economies, 12(12), 338.

Pili, S., Serra, P., & Salvati, L. (2019). Landscape and the city: Agro-forest systems, land fragmentation and the ecological network in Rome, Italy. Urban Forestry & Urban Greening, 41, 230–237.

Quaranta, G., Salvia, R., Cudlin, P., & Salvati, L. (2023). Evaluating the spatial coherence of composite indexes of land degradation at small administrative units. Environmental Impact Assessment Review, 103, 107226.

Quaranta, G., Salvia, R., Salvati, L., Paola, V. De, Coluzzi, R., Imbrenda, V., & Simoniello, T. (2020). Long-term impacts of grazing management on land degradation in a rural community of Southern Italy: Depopulation matters. Land Degradation & Development, 31(16), 2379–2394.

Ramsfield, T. D., Bentz, B. J., Faccoli, M., Jactel, H., & Brockerhoff, E. G. (2016). Forest health in a changing world: effects of globalization and climate change on forest insect and pathogen impacts. Forestry, 89(3), 245–252.

Randin, C. F., Ashcroft, M. B., Bolliger, J., Cavender-Bares, J., Coops, N. C., Dullinger, S., Dirnböck, T., Eckert, S., Ellis, E., Fernández, N., Giuliani, G., Guisan, A., Jetz, W., Joost, S., Karger, D., Lembrechts, J., Lenoir, J., Luoto, M., Morin, X., … Payne, D. (2020). Monitoring biodiversity in the Anthropocene using remote sensing in species distribution models. Remote Sensing of Environment, 239, 111626.

Raymond, C. M., Bryan, B. A., MacDonald, D. H., Cast, A., Strathearn, S., Grandgirard, A., & Kalivas, T. (2009). Mapping community values for natural capital and ecosystem services. Ecological Economics, 68(5), 1301–1315.

Rivadeneyra, P., Scaccia, L., & Salvati, L. (2023). A spatial regression analysis of Colombia's narcodeforestation with factor decomposition of multiple predictors. Scientific Reports, 13(1), 13485.

Rodrigo-Comino, J., Salvia, R., Quaranta, G., Cudlín, P., Salvati, L., & Gimenez-Morera, A. (2021). Climate Aridity and the Geographical Shift of Olive Trees in a Mediterranean Northern Region. Climate, 9(4), 64.

Rosa, E. A., & Dietz, T. (1998). Climate Change and Society. International Sociology, 13(4), 421–455.

Rosa, E. A., York, R., & Dietz, T. (2004). Tracking the Anthropogenic Drivers of Ecological Impacts. Ambio: A Journal of the Human Environment, 33(8), 509–512.

Roy, P. S., & Tomar, S. (2000). Biodiversity characterization at landscape level using geospatial modelling technique. Biological Conservation, 95(1), 95–109.

Sallustio, L., Pettenella, D., Merlini, P., Romano, R., Salvati, L., Marchetti, M., & Corona, P. (2018). Assessing the economic marginality of agricultural lands in Italy to support land use planning. Land Use Policy, 76, 526–534.
Salvati, L. (2013). Monitoring high-quality soil consumption driven by urban pressure in a growing city (Rome, Italy). Cities, 31, 349–356.
Salvati, L. (2014). Profiling forest fires along the urban gradient: a Mediterranean case study. Urban Ecosystems, 17(4), 1175–1189.
Salvati, L. (2023). Two wrongs don't make a right: A multi-step decomposition of latent dimensions of sustainable development and desertification risk in Italy. Ecological Economics, 212, 107926.
Salvati, L., & Carlucci, M. (2014). Zero Net Land Degradation in Italy: The role of socioeconomic and agro-forest factors. Journal of Environmental Management, 145, 299–306.
Salvati, L., & Carlucci, M. (2015). Towards sustainability in agro-forest systems? Grazing intensity, soil degradation and the socioeconomic profile of rural communities in Italy. Ecological Economics, 112, 1–13.
Salvati, L., & Ferrara, A. (2014). Do land cover changes shape sensitivity to forest fires in peri-urban areas? Urban Forestry & Urban Greening, 13(3), 571–575.
Salvati, L., & Ferrara, A. (2015). Validation of MEDALUS Fire Risk Index using Forest Fires Statistics through a multivariate approach. Ecological Indicators, 48, 365–369.
Salvati, L., Ferrara, A., & Chelli, F. (2018). Long-term growth and metropolitan spatial structures: an analysis of factors influencing urban patch size under different economic cycles. Geografisk Tidsskrift-Danish Journal of Geography, 118(1), 56–71.
Salvati, L., Mavrakis, A., Colantoni, A., Mancino, G., & Ferrara, A. (2015). Complex Adaptive Systems, soil degradation and land sensitivity to desertification: A multivariate assessment of Italian agro-forest landscape. Science of The Total Environment, 521–522, 235–245.
Salvati, L., & Ranalli, F. (2015). 'Land of Fires': Urban growth, economic crisis, and forest fires in Attica, Greece. Geographical Research, 53(1), 68–80.
Salvati, L., Ranalli, F., Carlucci, M., Ippolito, A., Ferrara, A., & Corona, P. (2017). Forest and the city: a multivariate analysis of peri-urban forest land cover patterns in 283 European metropolitan areas. Ecological Indicators, 73, 369–377.
Salvati, L., Sabbi, A., Smiraglia, D., & Zitti, M. (2014). Does forest expansion mitigate the risk of desertification? Exploring soil degradation and

land-use changes in a Mediterranean country. International Forestry Review, 16(4), 485–496.

Salvati, L., & Zitti, M. (2009). Substitutability and weighting of ecological and economic indicators: Exploring the importance of various components of a synthetic index. Ecological Economics, 68(4), 1093–1099.

Salvia, R., Quaranta, G., Rontos, K., Cudlin, P., & Salvati, L. (2023). Investigating metropolitan hierarchies through a spatially explicit (local) approach. ISPRS International Journal of Geo-Information, 12(8), 315.

Samela, C., Imbrenda, V., Coluzzi, R., Pace, L., Simoniello, T., & Lanfredi, M. (2022). Multi-Decadal Assessment of Soil Loss in a Mediterranean Region Characterized by Contrasting Local Climates. Land, 11(7), 1010.

Seifollahi-Aghmiuni, S., Kalantari, Z., Egidi, G., Gaburova, L., & Salvati, L. (2022). Urbanisation-driven land degradation and socioeconomic challenges in peri-urban areas: Insights from Southern Europe. Ambio, 51(6), 1446–1458.

Smiraglia, D., Ceccarelli, T., Bajocco, S., Salvati, L., & Perini, L. (2016). Linking trajectories of land change, land degradation processes and ecosystem services. Environmental Research, 147, 590–600.

Treweek, J. (1996). Ecology and Environmental Impact Assessment. The Journal of Applied Ecology, 33(2), 191.

Turner, W., Spector, S., Gardiner, N., Fladeland, M., Sterling, E., & Steininger, M. (2003). Remote sensing for biodiversity science and conservation. Trends in Ecology & Evolution, 18(6), 306–314.

Usman, M., & Hammar, N. (2021). Dynamic relationship between technological innovations, financial development, renewable energy, and ecological footprint: fresh insights based on the STIRPAT model for Asia Pacific Economic Cooperation countries. Environmental Science and Pollution Research, 28(12), 15519–15536.

Vardopoulos, I., Falireas, S., Konstantopoulos, I., Kaliora, E., & Theodoropoulou, E. (2018). Sustainability assessment of the agri-environmental practices in Greece. Indicators' comparative study. International Journal of Agricultural Resources, Governance and Ecology, 14(4), 368.

Vardopoulos, I., Ioannides, S., Georgiou, M., Voukkali, I., Salvati, L., & Doukas, Y. E. (2023). Shaping sustainable cities: A long-term GIS-emanated spatial analysis of settlement growth and planning in a coastal Mediterranean European city. Sustainability, 15(14), 11202.

Westoby, M., Walker, B., & Noy-Meir, I. (1989). Opportunistic Management for Rangelands Not at Equilibrium. Journal of Range Management, 42(4), 266.

York, R., Rosa, E. A., & Dietz, T. (2003). STIRPAT, IPAT and ImPACT: analytic tools for unpacking the driving forces of environmental impacts. Ecological Economics, 46(3), 351–365.

Zambon, I., Colantoni, A., Carlucci, M., Morrow, N., Sateriano, A., & Salvati, L. (2017). Land quality, sustainable development and environmental degradation in agricultural districts: a computational approach based on entropy indexes. Environmental Impact Assessment Review, 64, 37–46.

3

Aesthetic Value, Biodiversity, and Ecosystem Services: Optimal Resources' Management and the "Holistic" Role of Landscape

Alessia D'Agata[1,2], Clio Ciaschini[3], Marco Maialetti[4], Donato Scarpitta[5], Rosanna Salvia[6], Giovanni Quaranta[6], and Ioannis Konaxis[7]

[1]Department of Methods and Models for Economics, Territory and Finance (MEMOTEF), Faculty of Economics, Sapienza University of Rome, Italy
[2]Department of Geography, Building B, Autonomous University of Barcelona, Spain
[3]Department of Economics and Social Sciences, Polytechnic University of Marche, Italy
[4]Independent Researcher and Lawyer, Italy
[5]Independent Researcher and Architect, Italy
[6]Department of Mathematics, Informatics and Economics, University of Basilicata, Italy
[7]Department of Tourism Studies, University of Piraeus, Greece

Email: alessia.dagata@uniroma1.it; c.ciaschini@univpm.it; avv.marcomaialetti@gmail.com; scarpitta.donato@libero.it; rosanna.salvia@unibas.it; giovanni.quaranta@unibas.it; ikonaxis@unipi.gr

Abstract

Climate change implies challenges for both natural ecosystems and human societies, necessitating prompt and coordinated mitigation and adaptation strategies. Understanding and managing these dynamics requires multi-dimensional approaches capable of integrating environmental, economic, and social knowledge. Current methods for quantifying environmental state, including natural capital and ecosystem services, provide essential

DOI: 10.1201/9788743811107-3

information but often fail to capture the resilience of local communities or the complex interactions between human and landscape systems. This contribution emphasizes the development of locally relevant indicators that bridge disciplinary boundaries, combining official statistics, geospatial data, and remote sensing with qualitative and historical knowledge. Such indicators are designed to inform sustainable land management, strategic spatial planning, and policy decisions aimed at mitigating climate impacts, particularly in natural habitats. By linking socioeconomic patterns, demographic dynamics, and environmental changes, this work contributes to a more nuanced understanding of spatial inequalities, ecological functionality, and the effects of human activities on ecosystem services. Ultimately, this approach supports evidence-based policy-making, facilitating adaptation and mitigation strategies aligned with sustainable development goals, while highlighting potential inequalities in resource access and environmental benefits. The operational framework proposed here emphasizes the importance of local-scale, holistic, and temporally informed indicators as a critical foundation for resilient and equitable ecosystem governance in the context of global warming.

3.1 Introduction

The well-known implications of climate change are far-reaching and have highlighted the urgency of taking prompt, decisive actions to mitigate and adapt to its impact. In order to monitor natural ecosystems, measure their economic value, or assess the potential for sustainable development of a given place, environmental policies currently depend on indicators derived from multiple knowledge frameworks. However, these methods of quantifying nature have limitations and do not take into account the resilience of local communities. In light of climate change, recent studies aim at implementing indicators that can offer a composite representation of the state of national ecosystems at a local scale, for example, with a focus on agricultural, natural, and forest ecosystems. These indicators can be novel in comparison with the computational methods currently in use, effectively bridging the gap between different disciplinary approaches and specialized knowledge, while improving the quality of statistical data on landscapes and forestry.

The first warning to humans about the need to reconsider their approach to environmental resources came in the early 1970s, prompted by seminal studies such as "The Entropy Law and the Economic Process" by Georgescu-Roegen in 1971, and "The Limits to Growth," as a result of research works led by Donella H. Meadows and Dennis L. Meadows in 1972. Both works emphasize the physical limits of economic growth. Consequently, economic thought has been deeply involved in this debate. The accumulation of physical and human capital was initially considered as an input variable to produce goods, services, and human welfare. Later on, natural capital and physical capital were better defined and made distinct (Pearce and Barbier, 2000).

An increased awareness of the "ecological" issue allowed a refined formulation of pivotal concepts, such as the hegemonic notion of "sustainable development" introduced in 1987 by the WCED (World Commission on Environment and Development) report entitled "Our Common Future," lately emphasizing the difference between efficiency and sustainability objectives (Bishop, 1993). The awareness of human responsibilities in global warming underscored the necessity for quick and effective action. Its effects, which are mostly caused by human activity, are extensive and affect many regions.

3.2 The Logical Background: A Brief Commentary

Caused by anthropogenic social structures that lead to greenhouse gas emissions from burning fossil fuels, deforestation, desertification, and various industrial and agricultural practices, climate change impacts the life of the earth, both for ecosystems and humans (e.g., health, food security, water resources, the economic system at large, migration, and governance), are profound and interlinked. Addressing these challenges requires urgent and coordinated (global) efforts to reduce emissions and adapt to changing conditions (Calvin et al., 2023).

Optimal land management and spatial planning are specifically required to manage the effects of ongoing climate change by improving land-use planning and urban design, emphasizing adaptation strategies like carbon sequestration and heat mitigation, and addressing risks like those associated with flooding (Storch and Downes, 2011; Runhaar et al., 2012; Bounoua et al., 2015). Furthermore, environmental and social challenges call for a revolution in land management techniques that incorporate environmental resources into frameworks aimed at sustainable (local) development.

The "holistic" feature of the "ecological" issue, driven by new social and demographic dynamics, requires optimal land planning to take a strategic approach that considers both global and local scales. Optimal resource management should especially base its arguments equally on the various development trajectories (e.g., economic, social, or environmental), the multiple geographies of local territories (namely, urban/peri-urban/rural), and complex knowledge frameworks inspecting human-environmental interactions (HEI), their evolution, and the related cause-and-effect mechanisms (Gibelli, 2003).

HEI refers to the interactions that take place between (human) social systems and the rest of the ecosystem. These are intricate adaptive systems due to their multiple components and interconnections, as well as their feedback mechanisms that support survival in a dynamic ecosystem or environment (Marten, 2001). To reconstruct these flows of human-environment interaction, the use of multidisciplinary sources and diverse knowledge is required (Gambino, 1997). It is around this knowledge that the information needed to formulate correct land management measures is concealed.

3.3 The Role of Official Statistics and Ecological Indicators

Environmental laws and policies rely on indicators developed from several knowledge frameworks. Ecological indicators are used to track environmental integrity, support sustainable strategies and development paths, and aid in informed decision-making. However, care should be given when assuming they are important parameters for environmental governance, and greater focus should be placed on how they are generated and applied in local contexts (Butt, 2018). A further (technical) problem for indication theory is the complex network of connections between environmental and socioeconomic issues. No one indicator can adequately address the whole range of equity issues that are pertinent to environmental policy (Bergh et al., 2001). In this regard, a multidisciplinary approach capable of considering the primary interactions between economic and biophysical dimensions could aid in understanding the socioeconomic processes that have an impact on the environment and the benefits that may result (Salvati, 2018). These assumptions highlight the need to incorporate information from distinctive academic fields when creating indicators to support official statistics, in order to facilitate policy making.

3.3.1 Measuring human activities

Environmental monitoring refers to those activities designed to verify, with a specific time schedule, the maintenance of the environmental conditions at a given ecosystem. In a nutshell, environmental monitoring is crucial to understanding and handling the effects of human activities on the environment, promoting a balance between economic development and ecosystem conservation. Monitoring is related to the impact of anthropogenic actions over the specified time interval, and the primary (or secondary) sources used to create the key environmental monitoring indicators might range widely. Official statistics provide measurements, sampling, or counting, and thus data from field monitoring activities, but they also offer data analysis to find trends, create distribution patterns, or forecast impacts. Aggregated data, such as information on the use of natural resources globally or regionally, can also be used to produce environmental indicators.

3.3.2 Pricing the environment: Natural capital and ecosystem services

If we discuss natural resources from the standpoint of their economic value and the services they provide, the term "natural capital" is frequently used (Costanza, 1997). This work has drawn criticism and discussion for several reasons, and economists and environmentalists dispute its application, which is seen as a threat to economic growth on the one hand, and a break from the concept of "biodiversity" on the other hand. Additionally, it is rather unclear if it is possible to put a meaningful (numerical) value quantifying benefits derived from natural features (Costanza and Daly, 1992; Edwards and Abivardi, 1998; Naidoo et al., 2009). An important approach to quantifying the value of natural capital has been provided by the concept of ecosystem services (Millennium Ecosystem Assessment (MEA), 2001).

Quantifying the economic value of ecosystem services provided by nature is a complex and debated topic, and there are obstacles to accurate measurement. However, quantifying the economic value of ecosystem services – which is crucial for informing policy and natural resource management decisions – requires a multidisciplinary approach, and several obstacles are encountered in its computation (Ferrara et al., 2017). Many of the benefits that ecosystems provide, for instance, are non-commercial or intangible (e.g., Sallustio et al., 2018). Purification of the air we breathe or the water we drink, or climate regulation and pollination, are examples of such benefits. In the absence of a reference market or an exchange price, they are not part of cost/benefit balances and are not considered in decision-making and intervention processes (Seifollahi-Aghmiuni et al., 2022).

Another obstacle is the existence value, which corresponds to what is intrinsic to a resource and does not correspond to something directly employed. This can be more difficult to examine because it is not easily quantifiable in economic terms. Finally, human activities can create pressures on ecosystems that affect their conservation status and capacity to provide ecosystem services. These pressures can have both positive and negative effects, but are often difficult to quantify in economic terms (Blasi et al., 2015).

3.3.3 Measuring sustainability

As part of the discussion on the value of ecosystems and socio-environmental indicators, some support is found within the 17 global Sustainable Development Goals (SDGs) enshrined by the United Nations in the 2030 Agenda in 2015. Among the 169 sub-goals, in fact, there are references to benchmarks to comply with regarding the natural elements. Computation strategies have been implemented, trying to link the biophysical factors of the natural elements to the social factors of local (and regional) development. However, the SDGs are often criticized for their breadth and lack of a clear hierarchy (or priority) among the goals, and this can lead to inherent difficulties in their implementation and dispersion of resources (Hickel, 2019).

In addition, the lack of reliable data and standardized methods for measuring progress toward the SDGs can be a significant concern, making it difficult to monitor actual progress and assess the impact of implemented policies, since not every country in the world is aligned with the official statistics requirements of the United Nations Statistical Commission. In addition, while containing goals addressing both the environmental sphere and the mitigation and adaptation of territories to the effects of climate change (Goals 13-15), there is a latent conflict between these targets and the goals of promoting economic growth (Goal 8). Some critical positions argue that the emphasis on economic growth could undermine efforts directed toward environmental sustainability (Hickel & Kallis, 2020). These critical issues obviously become more intense as we scale down and consider work at a local scale.

3.3.4 Operational approaches

Quantification of resources specifically aimed at describing the state (or changes) in natural or human ecosystems – measuring their economic value or the capacity for sustainable development of a given area – has critical issues. Individual approaches usually lack the ability to indicate the degree of resilience of local communities (Salvati & Zitti, 2008). Monitoring indicators have potential but often fail to materialize because they are not applicable at all scales, having in turn little interaction with disciplines outside the environmental world. In giving socioeconomic indications of nature, even talking about natural capital and ecosystem services, is neither functional nor conceptually conducive to nature conservation policies (Salvati & Zitti, 2009; Salvati & Carlucci, 2014; Tomao et al., 2017).

Monetary valuation, intrinsic to the concept of natural capital, reduces nature to a commodity, ignoring its non-economic values, and defining nature as capital and its services as products (Farley, 2012). This vision paves the way for their commodification and possible exploitation for profit. The SDGs, in addition to suggesting critical issues regarding the methods of computing indicators at the local scale and the sources used, hold inherent inconsistency and difficult dialogue between the goals of environmental protection and economic growth (Hickel, 2019).

Based on these premises, we suggest adopting a multidimensional approach that recognizes the plurality of natural values and promotes their long-term protection through institutional arrangements and policies that go beyond the logic of profit or simple economic growth. More specifically, we aim at studying and developing indicator systems capable of providing information at the local scale that can illustrate the complexity of the relationships between the environmental and socioeconomic systems, making it capable of informing adaptation and mitigation policies, as well as directing sustainable development paths (Salvati & Carlucci, 2014). The focus, as far as the environmental sphere is concerned, is basically on natural habitats, because of their importance in mitigating the effects of climate change (Salvati & Zitti, 2005; Ferrara et al., 2016; Mancino et al., 2016).

Forestry is one of the environmental dimensions for which official national statistics are rather weak, both economically and ecologically. For instance, there is a lack of data on the value of forests from a biophysical and, at the same time, economic perspective, and of forestry activities from a national accounting perspective. Today, discontinuous surveys do not make it possible to obtain a complete picture of the state of evolution of forests, and for this topic, the strong interest is not only economic, but also ecological, due to the vast ensemble of benefits that forests offer and that affect territorial stakeholders.

3.3.5 The role of geospatial data

When it comes to monitoring concerns associated with land use-cover change, satellite data is one of the primary sources for indicators since it may offer extensive information on a variety of environmental phenomena, such as the usage of natural resources or the quality of air and water (Nabuurs et al., 2022). By combining the design and deployment of data acquisition equipment with data processing/interpretation processes, remote sensing acquires information about a field object (or phenomenon) from an acquisition instrument that is not in direct contact with the object under investigation. So, data on complex socioeconomic phenomena and spatial dynamics of landscape transformations, such as land consumption and degradation,

Figure 3.1 Man-made green landscapes between urban and rural districts. Source: authors' photographic archive.

will be associated with environmental data produced and validated from earth observation sources (Salvati, 2013; Francaviglia et al., 2019; Ferrara et al., 2020).

The information gathered and the methodology used to process are eminently multidisciplinary – encompassing statistics, spatial econometrics, and remote sensing. In addition, the approaches based on remote sensing address the relationship between socioeconomic and environmental impacts, so that measurement can be more effectively reproduced even at a local scale, where land use changes drastically (Incerti et al., 2007). A consistency analysis of statistical indices based on satellite observation can be finally implemented using advanced statistical methodologies. A comparative analysis of existing indicators will finally stress the importance of already available administrative information that is difficult to compare in time and space in other ways (Ferretti et al., 2024). Analyzing the effect of current land-use governance and conservation policies (Figure 3.1), such as the establishment of protected areas, allows an indirect test of the effectiveness of environmental guidance to be implemented within land-use planning and protection processes (Salvati et al., 2012).

3.4 Concluding Remarks

Given the need to build up multi-disciplinary knowledge frameworks to produce indicators prior to the formulation of strategic environmental management plans, it is important to move toward the analysis of quantitative information to propose a significant improvement to the current data heritage. In this perspective, an operational proposal aimed at developing indicators provides a composite representation of ecosystem state at a local scale, in the light of climate change. Up to now, empirical studies have successfully delineated the gap between various disciplines and specific information, while fixing the partiality of statistical data regarding the "ecological" sphere (Colantoni et al., 2015).

The analysis of statistical data on environmental elements, specifically forests, and the development of holistic land-use indicators are required for the implementation of multidisciplinary frameworks providing policy suggestions and informing strategic planning oriented toward sustainable development (Salvati & Carlucci, 2011). In the context of climate change, such plans are even more important since they are required for adaptation and mitigation to its consequences. Furthermore, because forest expansion is one of the primary (formal and informal) operational targets for mitigating the effects of climate change, particularly global warming, improved forestry information implies the ability to detect spatial inequalities in resource supply, and thus, of related benefits or vulnerabilities, indirectly addressing the issue of climate inequalities.

References

Bishop, R. C. (1993). Economic Efficiency, Sustainability, and Biodiversity. Ambio, 22(2/3), 69–73.

Blasi, C., Capotorti, G., Alós Ortí, M. M., Anzellotti, I., Attorre, F., Azzella, M. M., Carli, E., Copiz, R., Garfì, V., Manes, F., Marando, F., Marchetti, M., Mollo, B., & Zavattero, L. (2017). Ecosystem mapping for the implementation of the European Biodiversity Strategy at the national level: The case of Italy. Environmental Science & Policy, 78, 173–184.

Carbone, J., & Smith, V. K. (2010). Valuing ecosystem services in general equilibrium (w15844; p. w15844). National Bureau of Economic Research.

Colantoni, A., Ferrara, C., Perini, L., & Salvati, L. (2015). Assessing trends in climate aridity and vulnerability to soil degradation in Italy. Ecological indicators, 48, 599–604.

Costanza, R., & Daly, H. E. (1992). Natural Capital and Sustainable Development. Conservation Biology, 6(1), 37–46.

Edwards, P. J., & Abivardi, C. (1998). The value of biodiversity: Where ecology and economy blend. Biological Conservation, 83(3), 239–246.

Farley, J. (2012). Ecosystem services: The economics debate. Ecosystem Services, 1(1), 40–49.

Ferrara, A., Kelly, C., Wilson, G. A., Nolè, A., Mancino, G., Bajocco, S., & Salvati, L. (2016). Shaping the role of 'fast' and 'slow' drivers of change in forest-shrubland socio-ecological systems. Journal of Environmental Management, 169, 155–166.

Ferrara, A., Kosmas, C., Salvati, L., Padula, A., Mancino, G., & Nolè, A. (2020). Updating the MEDALUS-ESA framework for worldwide land degradation and desertification assessment. Land Degradation & Development, 31(12), 1593–1607.

Ferrara, C., Carlucci, M., Grigoriadis, E., Corona, P., & Salvati, L. (2017). A comprehensive insight into the geography of forest cover in Italy: Exploring the importance of socioeconomic local contexts. Forest Policy and Economics, 75, 12–22.

Ferretti, M., Fischer, C., Gessler, A., Graham, C., Meusburger, K., Abegg, M., Bebi, P., Bergamini, A., Brockerhoff, E. G., Brunner, I., Bühler, C., Conedera, M., Cothereau, P., D'Odorico, P., Düggelin, C., Ginzler, C., Grendelmeier, A., Haeni, M., Hagedorn, F., Shackleton, R. T. (2024). Advancing forest inventorying and monitoring. Annals of Forest Science, 81(1), 6.

Follett, R., & Strezov, V. (2015). An Analysis of Citizen Science Based Research: Usage and Publication Patterns. Plos One, 10(11), e0143687.

Francaviglia, R., Di Bene, C., Farina, R., Salvati, L., & Vicente-Vicente, J. L. (2019). Assessing "4 per 1000" soil organic carbon storage rates under Mediterranean climate: a comprehensive data analysis. Mitigation and Adaptation Strategies for Global Change, 24, 795–818.

Goodchild, M. F. (2007). Citizens as sensors: The world of volunteered geography. GeoJournal, 69(4), 211–221.

Hickel, J. (2019). The contradiction of the sustainable development goals: Growth versus ecology on a finite planet. Sustainable Development, 27(5), 873–884.

Hickel, J., & Kallis, G. (2020). Is Green Growth Possible? New Political Economy, 25(4), 469–486.

Incerti, G., Feoli, E., Salvati, L., Brunetti, A., & Giovacchini, A. (2007). Analysis of bioclimatic time series and their neural network-based

classification to characterise drought risk patterns in South Italy. International Journal of Biometeorology, 51, 253–263.
Jollands, N. (2006). How to aggregate sustainable development indicators: A proposed framework and its application. International Journal of Agricultural Resources, Governance and Ecology, 5(1), 18.
Mancino, G., Nolè, A., Salvati, L., & Ferrara, A. (2016). In-between forest expansion and cropland decline: A revised USLE model for soil erosion risk under land-use change in a Mediterranean region. Ecological Indicators, 71, 544–550.
Marten, G. G. (2001). Human ecology: Basic concepts for sustainable development. Earthscan, London.
Meadows, D. H., Club of Rome, & Potomac Associates. (1972). The limits to growth. Earth Island, New York.
Millennium Ecosystem Assessment (2005). Ecosystems and human well-being: Synthesis. Island Press, New York.
Nabuurs, G.-J., Harris, N., Sheil, D., Palahi, M., Chirici, G., Boissière, M., Fay, C., Reiche, J., & Valbuena, R. (2022). Glasgow forest declaration needs new modes of data ownership. Nature Climate Change, 12(5), 415–417.
Naidoo, R., Malcolm, T., & Tomasek, A. (2009). Economic benefits of standing forests in highland areas of Borneo: Quantification and policy impacts. Conservation Letters, 2(1), 36–45.
Pearce, D., & Barbier, E. (2000). Blueprint for a Sustainable Economy. Routledge, London.
Runhaar, H., Mees, H., Wardekker, A., Van Der Sluijs, J., & Driessen, P. P. J. (2012). Adaptation to climate change-related risks in Dutch urban areas: Stimuli and barriers. Regional Environmental Change, 12(4), 777–790.
Sallustio, L., Pettenella, D., Merlini, P., Romano, R., Salvati, L., Marchetti, M., & Corona, P. (2018). Assessing the economic marginality of agricultural lands in Italy to support land use planning. Land Use Policy, 76, 526–534.
Salvati, L. (2013). Monitoring high-quality soil consumption driven by urban pressure in a growing city (Rome, Italy). Cities, 31, 349–356.
Salvati, L., & Carlucci, M. (2011). The economic and environmental performances of rural districts in Italy: Are competitiveness and sustainability compatible targets? Ecological Economics, 70(12), 2446–2453.
Salvati, L., & Carlucci, M. (2014). A composite index of sustainable development at the local scale: Italy as a case study. Ecological Indicators, 43, 162–171.

Salvati, L., & Zitti, M. (2005). Land degradation in the Mediterranean basin: Linking bio-physical and economic factors into an ecological perspective. Biota, 6, 67–77.

Salvati, L., & Zitti, M. (2008). Regional convergence of environmental variables: empirical evidences from land degradation. Ecological Economics, 68(1-2), 162–168.

Salvati, L., & Zitti, M. (2009). Substitutability and weighting of ecological and economic indicators: Exploring the importance of various components of a synthetic index. Ecological Economics, 68(4), 1093–1099.

Salvati, L., Perini, L., Sabbi, A., & Bajocco, S. (2012). Climate aridity and land use changes: A regional-scale analysis. Geographical Research, 50(2), 193–203.

Seifollahi-Aghmiuni, S., Kalantari, Z., Egidi, G., Gaburova, L., & Salvati, L. (2022). Urbanisation-driven land degradation and socioeconomic challenges in peri-urban areas: Insights from Southern Europe. Ambio, 51(6), 1446–1458.

Storch, H., & Downes, N. K. (2011). A scenario-based approach to assess Ho Chi Minh City's urban development strategies against the impact of climate change. Cities, 28(6), 517–526.

Tomao, A., Quatrini, V., Corona, P., Ferrara, A., Lafortezza, R., & Salvati, L. (2017). Resilient landscapes in Mediterranean urban areas: Understanding factors influencing forest trends. Environmental Research, 156, 1–9.

World Commission on Environment and Development (1987). Our common future. Oxford University Press, Oxford.

4

Investigating Complex Landscape Systems: Inter-disciplinary Paradigms In-between Science and Policy

Adele Sateriano[1], Alessia D'Agata[2,3], Marco Maialetti[4], Donato Scarpitta[5], Andrea Colantoni[6], Luca Salvati[2], and Ioannis Konaxis[7]

[1]Mediterranean Foundation for Sustainable Development (MEDES), Italy
[2]Department of Methods and Models for Economics, Territory and Finance (MEMOTEF), Faculty of Economics, Sapienza University of Rome, Italy
[3]Department of Geography, Building B, Autonomous University of Barcelona, Spain
[4]Independent Researcher and Lawyer, Italy
[5]Independent Researcher and Architect, Italy
[6]Department of Agricultural and Forest Sciences (DAFNE), University of Tuscia, Italy
[7]Department of Tourism Studies, University of Piraeus, Greece

Email: adele@medes.eu; alessia.dagata@uniroma1.it; avv.marcomaialetti@gmail.com; scarpitta.donato@libero.it; colantoni@unitus.it; luca.salvati@uniroma1.it; ikonaxis@unipi.gr

Abstract

Mediterranean-like ecosystems host one of the world's most significant biodiversity hotspots, where traditional land-use practices have historically shaped resilient socio-ecological systems. Among these, terraced landscapes represent complex systems that stabilize steep slopes, support agricultural production, and provide diverse and multiple ecosystem services while reflecting centuries of accumulated local knowledge. Despite their cultural, ecological, and economic value, many terraced landscapes are threatened by

DOI: 10.1201/9788743811107-4

abandonment, land degradation, and socioeconomic marginalization, being in turn are compounded by climate change and rural depopulation. Sustainable management of these landscapes requires a comprehensive understanding of their ecological functions, socioeconomic dimensions, and historical transformations. Integrative monitoring approaches combining multi-temporal datasets, remote sensing, field surveys, and geographic information systems enable the evaluation of biodiversity, ecosystem services, land-use dynamics, and traditional ecological knowledge. Such assessments can inform strategies for terrace stabilization, restoration of degraded areas, and adaptation of traditional practices to contemporary socio-environmental conditions. Holistic analyses of terraced landscapes emphasize their multifunctional role in promoting biodiversity conservation, supporting agro-forestry and multifunctional agriculture, and fostering socio-cultural and economic resilience in rural areas. By integrating ecological, agricultural, and socioeconomic perspectives, these frameworks facilitate informed decision-making, landscape planning, and policy development. Preserving and revitalizing terraced systems not only safeguards cultural heritage but also enhances ecological stability, supports local livelihoods, and contributes to sustainable rural development in the context of global environmental change.

4.1 Introduction

The Mediterranean region is one of the 25 worldwide biodiversity hotspots (Myers et al., 2000). According to Naveh (1994, 1995), Mediterranean landscapes have been formed through a land-use sedimentation lasting thousands of years, having developed manifold resilient systems taken as a reflection of continuous anthropogenic impacts (Médail & Quézel, 1999). However, many regions with their traditional land-use systems, in particular mountain areas, underwent abandonment and degradation of long-lasting resilient agricultural systems (Serra et al., 2008; Van Der Sluis et al., 2014; Otero et al., 2015).

A unique example of these complex socio-ecological systems is represented by terraced systems, which have stabilized not only the socioeconomic conditions of mountain districts but also local environmental conditions, in particular steep slopes (Petit et al., 2012, Jiao et al., 2014; Arnáez et al., 2015, Špulerovà et al., 2017). By highlighting the long-lasting historical relationship between man and his territory, traditional knowledge in terrace landscapes has evolved, allowing a deeper understanding of the ecological and socioeconomic systems, providing in turn a basis for deriving future sustainable land-use scenarios, in particular when combined with modern land-use approaches and techniques (McCarter et al., 2014).

The high value of these anthropogenic landscapes has been recognized on the national and international level, for example, by preserving them as nature conservation areas (Berkes, 2012).

Terraced landscapes represent a complex anthropogenic system with manifold ecosystem services for the local communities and the society (e.g., Agnoletti et al., 2011), and have to be considered as an important cultural heritage (Daniel et al., 2012). In many cases, these include agro-forest systems with crop production and fruit trees on terraces, constructed with and stabilized by walls. These vertical landscapes can be found, besides many other parts of the Mediterranean region, in mountainous regions along latitudinal and longitudinal gradients, from the sea coast to the very inner land. They occur in remote, rural areas, as well as in peri-urban areas, and represent one of the oldest and most widespread sustainable forms of traditional landscape management. They are frequently associated with highly specialized agro-cultural adaptations that sum up into a complex traditional knowledge system (Modica et al., 2017). Terraced landscapes have often persisted over time, giving rise to distinctly characterized agricultural production that has found its way into the more "evolved" market thanks to the added value of their cultural and environmental premises.

4.2 Terraced Landscapes under Rising Human Pressure

Despite the international interest in the protection of terraced landscapes (e.g., Agnoletti, 2006, 2014), and apart from place-specific interventions, there is still no unifying strategic plan, which is based on holistic, multi-disciplinary knowledge and regional inventories. Integrative management plans identifying best practices are increasingly required (Psychogiopoulou, 2008). Therefore, a sustainable future of those vertical landscapes claims a comprehensive inventory, a detailed monitoring approach, and a holistic understanding of the relationships between the ecological system and the socioeconomic dimension (Ferrara et al., 2016). Against this background, the knowledge of the current distribution of terrace landscapes and the main changes occurred over the past decades is strategic (Incerti et al., 2007; Colantoni et al., 2015; Salvati et al., 2015).

On the one hand, it improves the overall knowledge of terrestrial land-use systems and main environmental and anthropogenic risk factors (Salvati & Carlucci, 2014). On the other hand, it allows for assessing the impact of local environmental conditions and management options on the future economic productivity and environmental sustainability of these systems and the definition of the best management and policy options (Tomao et al., 2017).

Furthermore, the identification of the most sensible areas that are endangered by multiple risk factors is considered a key element for addressing priorities, as well as optimizing conservation or restoration strategies (Biasi et al., 2015, 2019).

Modern conservation strategies for terraces should be based on (i) sustainable, traditional agro-cultural land-use systems, as well as on (ii) the protection of relict ecosystems with a high natural value, according to the balance of socio-ecological systems with high crop variety and natural species diversification (Salvati et al., 2013). Terraced landscapes have experienced landscape fragmentation, economic marginalization, and an increasing rate of abandonment, resulting in land instability and hydrogeological hazard (Arnáez et al., 2015). Therefore, the process of global change poses pressures upon these systems, including climate change with prolonged drought periods and extreme weather events, as well as the continuous depopulation of rural areas and the consequent loss of the associated traditional knowledge (Salvati et al., 2012; Stanchi et al., 2012; Mancino et al., 2016).

In these regards, there is a growing demand for agricultural short-chain production (Salvati & Zitti, 2005). This can be offered by these terraced landscapes to urban markets based on a new rural development strategy which is in accordance with the recognition of the value of traditional farming practices such as the "Intangible Cultural Heritage" formulated, among others, by UNESCO, and the role of local communities in the preservation of natural and socioeconomic resources of rural areas (Ferrara et al., 2016, 2017). In accordance with the objectives of environmental protection, the promotion of terraced landscapes plays a role in the conservation of biodiversity, the development of multifunctional agriculture, as well as agroforestry, the strengthening of local and regional economy, and a new relationship between metropolitan agglomerations and the countryside (Tomao et al., 2017). Consequently, maintenance and reactivation of these traditional land-use systems offer a sustainable solution related to the problems of urbanization and land abandonment (Quaranta et al., 2020).

4.3 Monitoring Approaches to Landscape Complexity

Country-wide inventories of terraced landscapes and a comprehensive monitoring system with information on ecosystems' services, land-use, and socioeconomic systems are increasingly required. Such surveillance systems should be based on multi-temporal datasets derived from different sources (remote sensing, field surveys, land cadaster, other indications from Geographic

Information Systems, aerial photographs, landscape photographs, as well as minor – or indirect – archival data). Implementation, calibration, and validation of a specific change detection model for terraced agricultural systems can be run using informatics and statistical tools (Günthert et al., 2011; Gran et al., 2013; Modica et al., 2017). This strategy should consider innovative multi-scalar assessments of agro-biodiversity that, besides ensuring typical and peculiar agro-ecological functions and services for the protection of abiotic and biotic resources – including the cultural identity of these landscapes – may also be suitable for predicting adaptation measures to climate change (Zaven, 2008).

The economic transition that characterizes these marginal areas justifies an integrated monitoring approach including environmental and socioeconomic aspects, which is perfectly in line with a holistic view of sustainability (Figure 4.1). These approaches are aimed at evaluating past, present, and future evolution of terraced landscapes, possibly identifying sustainable development scenarios (Sallustio et al., 2018). The main direct and indirect factors of change can be delineated through in-depth field investigations in pilot areas to promote appropriate conservation and promotion strategies from the perspective of a truly sustainable, integrative, and ecologically efficient rural development and preservation of cultural heritage (Galanina et al., 2017).

Based on fine-scale analysis, different scenarios can be delineated for the future evolution of such landscapes, including: (i) maintenance and re-activation of traditional land-use systems on terraces, modified by modern practices of agro-forestry; (ii) restoration and development of stabilized, near-natural ecosystems in those cases where a re-activation of terrace landscapes is not possible in the long-term; as well as (iii) stabilization of terraces with the future option to re-activate the traditional land-use system when socioeconomic conditions will evolve (e.g., Kosmas et al., 2015).

Despite the fact that the practice of ecosystem restoration has developed rapidly in the past decades (e.g., Van Andel & Aronson, 2012), there is a lack of knowledge of how to stabilize slopes with terraces after they are abandoned and thus prevent natural hazards (e.g., Carl & Richter, 1989; Tarolli et al., 2014; Arnaez et al., 2015). On the other hand, new terracing systems promoted by the European Union (EU) Common Agricultural Policy based on heavy earth movement, have been proved to have a highly negative landscape impact, turning out in material displacement, loss of the original soil profile and, when realized in the absence of traditional constructing practices, increase the risk of landslides (Martínez-Casasnovas et al., 2010).

Figure 4.1 Mixed rural landscapes in economically disadvantaged districts in Southern Europe; source: authors' photographic archive.

4.4 Concluding Remarks

Information systems supporting decisions in the field of sustainable management of terraces may stimulate a shared dissemination of best practices to stakeholders that operate in territorial contexts with similar background characteristics. Development of a common methodology for data collection and provision definitely produces a common geographical knowledge with fully shared and comparable environmental information. Building up a decision support system contributes to enhanced dissemination of empirical results derived from environmental research and surveillance, using interactive maps, datasets, and geo-localized information for the needs of final users.

References

Agnoletti, M. (2006). The conservation of cultural landscapes. CABI.

Agnoletti, M. (2014). Rural landscape, nature conservation and culture: Some notes on research trends and management approaches from a (Southern) European perspective. Landscape and Urban Planning, 126, 66–73.

Agnoletti, M., Cargnello, G., Gardin, L., Santoro, A., Bazzoffi, P., Sansone, L., Pezza, L., & Belfiore, N. (2011). Traditional landscape and rural development: Comparative study in three terraced areas in northern, central and southern Italy to evaluate the efficacy of GAEC standard 4.4 of cross compliance. Italian Journal of Agronomy, 6(1).

Arnáez, J. R., Lana-Renault, N., Lasanta, T., Ruiz-Flaño, P., & Castroviejo, J. (2015). Effects of farming terraces on hydrological and geomorphological processes: A review. Catena, 128, 122–134.

Barbera, G., & Cullotta, S. (2012). An inventory approach to the assessment of main traditional landscapes in Sicily (Central Mediterranean Basin). Landscape Research, 37(5).

Berkes, F. (2012). Sacred ecology: Traditional ecological knowledge and resource management. Taylor & Francis, London.

Biasi, R., Brunori, E., Ferrara, C., & Salvati, L. (2019). Assessing impacts of climate change on phenology and quality traits of Vitis vinifera L.: the contribution of local knowledge. Plants, 8(5), 121.

Biasi, R., Brunori, E., Smiraglia, D., & Salvati, L. (2015). Linking traditional tree-crop landscapes and agro-biodiversity in Central Italy using a database of typical and traditional products: A multiple risk assessment through a data mining analysis. Biodiversity and Conservation, 24(12), 3009–3031.

Boulangeat, I., Philippe, P., Abdulhak, S., Douzet, R., Garraud, L., Lavergne, S., Lavorel, S., Van Es, J., Vittoz, P., & Thuiller, W. (2012). Improving plant functional groups for dynamic models of biodiversity: At the crossroads between functional and community ecology. Global Change Biology, 18(11).

Carl, T., & Richter, M. (1989). Geoecological and morphological processes on abandoned vine-terraces in the Cinque Terre (Liguria). Geoökodynamik, 10, 125–158.

Colantoni, A., Ferrara, C., Perini, L., & Salvati, L. (2015). Assessing trends in climate aridity and vulnerability to soil degradation in Italy. Ecological Indicators, 48, 599–604.

Daniel, T. C., Muhar, A., Arnberger, A., Aznar, O., Boyd, J. W., Chan, K. M. A., Costanza, R., Elmqvist, T., Flint, C. G., Gobster, P. H., Grêt-Regamey, A., Lave, R., Muhar, S., Penker, M., Ribe, R. G., Schauppenlehner, T., Sikor, T., Soloviy, I., Spierenburg, M., & von der Dunk, A. (2012). Contributions of cultural services to the ecosystem services agenda. Proceedings of the National Academy of Sciences, 109(23), 8812–8819.

Ferrara, A., Kelly, C., Wilson, G. A., Nolè, A., Mancino, G., Bajocco, S., & Salvati, L. (2016). Shaping the role of 'fast' and 'slow' drivers of change in forest-shrubland socio-ecological systems. Journal of Environmental Management, 169, 155–166.

Ferrara, C., Carlucci, M., Grigoriadis, E., Corona, P., & Salvati, L. (2017). A comprehensive insight into the geography of forest cover in Italy: Exploring the importance of socioeconomic local contexts. Forest Policy and Economics, 75, 12–22.

Galanina, T. V., Baumgarten, M. I., Mikhailov, V. G., Koroleva, T. G., & Mikhailov, G. S. (2017). Environmental-socio-economic monitoring as a tool of region's environmental-economic system management. IOP Conference Series: Earth and Environmental Science, 50, 012030.

Georgiadis, T., Barbera, G., Carotenuto, F., Lenzi, J., Motisi, A., & Rossi, F. (2014). How a traditional agricultural protection structure acts in conditioning the internal microclimate: A statistical analytical approach to Giardino Pantesco (Pantelleria Island, Italy). Italian Journal of Agrometeorology, 19, 41–58.

Gran, K. B., Finnegan, N., Johnson, A. L., Belmont, P., Wittkop, C., & Rittenour, T. (2013). Landscape evolution, valley excavation, and terrace development following abrupt postglacial base-level fall. GSA Bulletin, 125(11-12), 1851–1864.

Günthert, S., Siegmund, A., Thunig, H., & Michel, U. (2011). Object-based detection of LUCC with special regard to agricultural abandonment on Tenerife (Canary Islands). Proceedings of SPIE - The International Society for Optical Engineering, 8181.

Incerti, G., Feoli, E., Salvati, L., Brunetti, A., & Giovacchini, A. (2007). Analysis of bioclimatic time series and their neural network-based classification to characterise drought risk patterns in South Italy. International Journal of Biometeorology, 51, 253–263.

Kosmas, C., Detsis, V., Karamesouti, M., Kounalaki, K., Vassiliou, P., & Salvati, L. (2015). Exploring long-term impact of grazing management on land degradation in the socio-ecological system of Asteroussia Mountains, Greece. Land, 4(3), 541–559.

Lavorel, S., & Garnier, E. (2002). Predicting changes in community composition and ecosystem functioning from plant traits: Revisiting the Holy Grail. Functional Ecology, 16, 545–556.

Mancino, G., Nolè, A., Salvati, L., & Ferrara, A. (2016). In-between forest expansion and cropland decline: A revised USLE model for soil erosion risk under land-use change in a Mediterranean region. Ecological Indicators, 71, 544–550.

Martínez-Casasnovas, J. A., Ramos, M. C., & Cots-Folch, R. (2010). Influence of the EU CAP on terrain morphology and vineyard cultivation in the Priorat region of NE Spain. Land Use Policy, 27(1), 11–21.

McCarter, J., Gavin, M. C., Baereleo, S., & Love, M. (2014). The challenges of maintaining indigenous ecological knowledge. Ecology and Society, 19(3), 39.

Médail, R., & Quézel, R. (1999). Biodiversity hotspots in the Mediterranean Basin: Setting global conservation priorities. Conservation Biology, 13, 1510–1513.

Modica, G., Praticò, S., & Di Fazio, S. (2017). Abandonment of traditional terraced landscape: A change detection approach (a case study in Costa Viola, Calabria, Italy). Land Degradation & Development, 28, 2608–2622.

Myers, N., Mittermeier, R. A., Mittermeier, C. G., da Fonseca, G. A. B., & Kent, J. (2000). Biodiversity hotspots for conservation priorities. Nature, 403, 853–858.

Naveh, Z. (1994). From biodiversity to ecodiversity: A landscape-ecology approach to conservation and restoration. Restoration Ecology, 2(3), 180–189.

Naveh, Z. (1995). Conservation, restoration, and research priorities for Mediterranean uplands threatened by global climate change. In: E. Pereira, C. Queiroz, H. Pereira, & L. Vicente (Eds.), Ecosystem services and human well-being: A participatory study in a mountain community in Portugal. Ecology and Society, 10(2), 14.

Petit, C., Konold, W., & Höchtl, F. (2012). Historic terraced vineyards: Impressive witnesses of vernacular architecture. Landscape History, 33(1).

Psychogiopoulou, E. (2008). Integration of cultural considerations in European Union law and policies. Martinus Nijhoff Publishers, Amsterdam.

Quaranta, G., Salvia, R., Salvati, L., Paola, V. D., Coluzzi, R., Imbrenda, V., & Simoniello, T. (2020). Long-term impacts of grazing management on land degradation in a rural community of Southern Italy: Depopulation matters. Land Degradation & Development, 31(16), 2379–2394.

Sallustio, L., Pettenella, D., Merlini, P., Romano, R., Salvati, L., Marchetti, M., & Corona, P. (2018). Assessing the economic marginality of agricultural lands in Italy to support land use planning. Land Use Policy, 76, 526–534.

Salvati, L., & Carlucci, M. (2014). A composite index of sustainable development at the local scale: Italy as a case study. Ecological Indicators, 43, 162–171.

Salvati, L., & Zitti, M. (2005). Land degradation in the Mediterranean basin: Linking bio-physical and economic factors into an ecological perspective. Biota, 6, 67–77.

Salvati, L., Mavrakis, A., Colantoni, A., Mancino, G., & Ferrara, A. (2015). Complex adaptive systems, soil degradation and land sensitivity to desertification: A multivariate assessment of Italian agro-forest landscape. Science of the Total Environment, 521, 235–245.

Salvati, L., Perini, L., Sabbi, A., & Bajocco, S. (2012). Climate aridity and land use changes: A regional-scale analysis. Geographical Research, 50(2), 193–203.

Salvati, L., Tombolini, I., Perini, L., & Ferrara, A. (2013). Landscape changes and environmental quality: the evolution of land vulnerability and potential resilience to degradation in Italy. Regional Environmental Change, 13, 1223–1233.

Špulerová, J., Dobrovodská, M., Štefunková, D., Kenderessy, P., & Izsóff, M. (2017). Terraced landscapes in Slovakia. Acta Geographica Slovenica, 57(2), 99–109.

Stanchi, S., Freppaz, M., Agnelli, A., Reinsch, T., & Zanini, E. (2012). Properties, best management practices and conservation of terraced soils in Southern Europe (from Mediterranean areas to the Alps): A review. Quaternary International, 265, 90–100.

Tomao, A., Quatrini, V., Corona, P., Ferrara, A., Lafortezza, R., & Salvati, L. (2017). Resilient landscapes in Mediterranean urban areas: Understanding factors influencing forest trends. Environmental Research, 156, 1–9.

van Andel, J., & Aronson, J. (2012). Restoration ecology: The new frontier. Wiley-Blackwell, London.

Wilson, G. A., Kelly, C. L., Briassoulis, H., Ferrara, A., Quaranta, G., Salvia, R., Detsis, V., Curfs, M., Cerda, A., El-Aich, A., Liu, H., Kosmas, C., Alados, C. L., Imeson, A., Landgrebe-Trinkunaite, R., Salvati, L., Naumann, S., Danwen, H., Iosifides, T., Kizos, T., Mancino, G., Nolè, A., Jiang, M., & Zhang, P. (2017). Social memory and the resilience of communities affected by land degradation. Land Degradation & Development, 28(2), 383–400.

Zeven, A. C. (1998). Landraces: A review of definitions and classifications. Euphytica, 104, 127–139.

5

Envisioning Landscape Structures with a Non-parametric, Multidimensional Analysis of Official Statistics

Clio Ciaschini[1], Donato Scarpitta[2], Alessia D'Agata[3,4], Enrico Maria Mosconi[5], Marco Maialetti[6], Adele Sateriano[7], and Ioannis Konaxis[8]

[1]Department of Economics and Social Sciences, Polytechnic University of Marche, Italy
[2]Independent Researcher and Architect, Italy
[3]Department of Methods and Models for Economics, Territory and Finance (MEMOTEF), Faculty of Economics, Italy
[4]Department of Geography, Building B, Autonomous University of Barcelona, Spain
[5]Department of Economics, Engineering, Society and Business Organization (DEIM), University of Tuscia, Italy
[6]Independent Researcher and Lawyer, Italy
[7]Mediterranean Foundation for Sustainable Development (MEDES), Italy
[8]Department of Tourism Studies, University of Piraeus, Greece

Email: c.ciaschini@univpm.it; scarpitta.donato@libero.it; alessia.dagata@uniroma1.it; enrico.mosconi@unitus.it; avv.marcomaialetti@gmail.com; adele@medes.eu; ikonaxis@unipi.gr

Abstract

This study presents a multidimensional framework for examining the composition of metropolitan landscapes through an extended form of principal component analysis (PCA) that incorporates alternative correlation metrics. Traditional PCA remains one of the most widely applied techniques in exploratory data analysis and predictive modelling, yet its reliance on linear

DOI: 10.1201/9788743811107-5

relationships and its sensitivity to outliers impose significant constraints when applied to complex spatial data. Metropolitan land-use systems are rarely governed by strictly linear associations, and the presence of outliers is often intrinsic to heterogeneous socioeconomic contexts. These limitations underscore the need for alternative approaches capable of revealing the latent structures embedded in large and multifaceted datasets. To address these methodological challenges, three PCA implementations were systematically compared using Pearson, Spearman, and Kendall correlation matrices. The empirical case study involved land-use data from several administrative units within the metropolitan region of Athens, Greece. The dataset describes the proportional distribution of a kaleidoscope of land-use categories, thus offering a high-resolution representation of spatial heterogeneity. By applying different correlation metrics, the study sought to determine how alternative conceptualizations of association influence dimensionality reduction and the interpretability of principal components. The results demonstrate that PCA based on Spearman's rank correlation provided a superior analytical performance in multiple respects. While Pearson's correlation accounted for the greatest share of variance in the first component, it proved less effective in subsequent dimensions, where Spearman consistently offered clearer discrimination among cases and more stable component loadings. The diagnostic indicators further reinforced this observation: both the Kaiser–Meyer–Olkin test and Bartlett's test suggested higher adequacy and stronger suitability for the Spearman-based solution. The robustness of Spearman's approach lies in its non-parametric nature, which makes it less vulnerable to the distortions caused by outliers and more capable of capturing monotonic, potentially non-linear relationships that characterize spatial and environmental data. Kendall's tau, while valuable in specific contexts, showed a more conservative explanatory capacity, highlighting that the choice of correlation metric directly affects the quality of dimensionality reduction.

5.1 Introduction

The analysis of large-scale datasets that integrate multiple statistical sources and spatial resolutions remains a central challenge in contemporary research, requiring continuous methodological innovation (Turner et al., 2007). Developing comprehensive approaches that encompass both spatial and temporal dimensions is crucial for uncovering latent structures within data distributions and for identifying key patterns that explain variability across complex systems (Brouwer et al., 1991). With the rapid expansion of available information across diverse disciplines (Conover, 1999), from the social sciences to environmental studies, researchers now face not only an unprecedented abundance of data but also the pressing need for exploratory techniques capable of addressing the challenges posed by non-normal distributions, non-linear relationships, and high levels of heterogeneity (Jolliffe & Cadima, 2016).

Among the multivariate tools available, principal component analysis (PCA) occupies a foundational role (Jolliffe, 2002). By transforming a set of correlated variables into a reduced number of uncorrelated components, PCA provides an effective means of summarizing complex relationships while minimizing information loss (Vyas & Kumaranayake, 2006). These components, derived through orthogonal transformations, are hierarchically ordered so that each successive axis captures the maximum possible variance remaining in the data (Benzécri, 1973). This structure allows for both a compact representation of multidimensional phenomena (Coppi & Bolasco, 1989). In practice, PCA has become a widely adopted technique across various domains, ranging from economics and sociology to ecology, chemistry, and physics, where it serves both exploratory and predictive purposes (Jolliffe & Cadima, 2016). The method yields two key outcomes: scores, which reposition observations within the new component space, and loadings, which indicate the relative influence of each original variable on the derived components (Johnson & Wichern, 2002).

As the scale and dimensionality of data increase, however, the classical formulation of PCA encounters important limitations. Its reliance on assumptions of linearity and sensitivity to outliers can compromise interpretability, particularly in datasets that are both large and structurally complex (Linton et al., 2007). For this reason, traditional PCA has progressively evolved into a broader family of generalizations, including non-linear, higher-order, robust, and weighted adaptations (Smarandache, 2009). Building on these advances, the present study introduces a non-parametric variant of PCA designed to capture non-linear associations more effectively while mitigating distortions caused by atypical values (Hollander & Wolfe, 1999). Specifically, three forms of PCA were applied to land-use data, each constructed on a different correlation matrix: the conventional Pearson matrix, and two non-parametric alternatives based on Spearman's (1904) rank and Kendall's (1970) tau (coefficients.

The choice of land-use data as a test case reflects its intrinsic complexity and relevance for contemporary spatial analysis (Ferrara et al., 2016). Land-use patterns emerge from the interplay of physical processes, ecological dynamics, and cultural practices, and their evolution reflects both long-term drivers and more rapid (short-term) transformations (Salvati & Zitti, 2005).

The resulting landscapes are characterized by intricate spatial configurations, ecological interdependencies, and structural variability that pose significant challenges to conventional statistical approaches (Turner et al., 2001). For this reason, exploratory multivariate methods should be adapted to handle non-linearities, non-normality, and other distributional irregularities that are basically common in such contexts (Wu & Hobbs, 2002). Non-parametric approaches, by relaxing the assumptions of linearity and normality, provide a valuable alternative for revealing hidden regularities and latent structures in large, heterogeneous datasets (Hollander & Wolfe, 1999). Their application offers not only methodological innovation but also practical insights for disciplines that increasingly rely on robust and adaptable tools to interpret the complexity of spatial and environmental systems (Turner et al., 2007).

5.2 Methodology

5.2.1 Logical framework

In the analysis of datasets comprising p variables observed across N cases, principal component analysis (PCA) simultaneously operates within two complementary spaces (variables and observations). The core procedure of PCA consists of the diagonalization of a symmetric matrix – either a covariance or a correlation matrix – depending on whether the data have been centered or standardized (Harman, 1976). These components collectively preserve the majority of the variance inherent in the dataset (Benzécri, 1973).

The transformation engenders a novel geometric framework, frequently referred to as the factor space, within which both observations and variables may be simultaneously represented and interpreted (Coppi & Bolasco, 1989). From a conceptual standpoint, PCA identifies optimal linear trajectories through the multidimensional cloud of data points, both in variable space and in observation space (Jolliffe, 2002). These trajectories are derived through least squares optimization, producing orthogonal components that maximize the variance explained within the dataset (Johnson & Wichern, 2002). The first principal component captures the largest proportion of variance, yet it seldom encompasses all variability; successive components are extracted to account for the remaining variation, always subject to the constraint that the total number of components cannot exceed the original number of variables.

This dual representation allows for the examination of both the structural relationships among variables and the distributional pattern of observations, providing a coherent framework for dimensionality reduction and multivariate interpretation (Jombart et al., 2008). By projecting high-dimensional data into a lower-dimensional factor space, PCA facilitates the identification of underlying patterns and latent structures, offering a robust basis for further exploratory or predictive analysis (Li et al., 2017).

5.2.2 Multidimensional analysis

In this study, principal component analysis (PCA) was conducted using correlation matrices rather than covariance matrices, in order to mitigate potential distortions arising from differences in variable scales. Traditionally, PCA employs the Pearson correlation matrix, which is derived from the Pearson product-moment correlation coefficient. This coefficient, ranging from −1 to +1, quantifies both the strength and direction of a linear relationship between two variables (Spearman, 1904). Values of +1 or −1 indicate a perfect positive or negative linear association, respectively, whereas a value of 0 means the absence of a linear relationship (Johnson & Wichern, 2002).

To extend the analytical scope beyond strictly linear associations, we incorporated two additional correlation matrices derived from non-parametric techniques, namely Spearman's rank correlation and Kendall's tau (Kendall, 1970). Both of these approaches assume that the variables under consideration are at least ordinal, allowing observations to be meaningfully ranked (Conover, 1999). Spearman's rank correlation evaluates how consistently one variable increases or decreases as a function of another, scrutinizing both linear and non-linear trends (Smarandache, 2009). In datasets without tied ranks, coefficients of +1 or −1 correspond to perfect monotonic relationships (Spearman, 1904). By operating on ranked values rather than raw measurements, Spearman's correlation exhibits greater robustness to outliers and deviations from normality compared to the Pearson coefficient (Hollander & Wolfe, 1999).

Kendall's tau provides an alternative rank-based measure of association, quantifying the degree to which the relative orderings of two variables are concordant (Kendall, 1970). Similar to Spearman's method, Kendall's tau is distribution-free, rendering it particularly suitable for non-parametric contexts and datasets where conventional assumptions of normality are not met. Leveraging these methodologies, we performed three distinct PCA analyses using correlation matrices constructed from Pearson, Spearman, and Kendall coefficients (Demšar et al., 2013). Subsequently, the resulting principal components were evaluated and compared through a series of diagnostic procedures, designed to assess their effectiveness and interpretability across differing correlation structures and data characteristics (Montano & Jombart, 2017).

5.2.3 Metrics and diagnostics

A central issue in the application of principal component analysis (PCA) lies in determining the number of principal components required to adequately summarize the full dataset, whether the focus is on variables or individual observations (Jolliffe & Cadima, 2016). Each eigenvalue derived from the correlation matrix quantifies the proportion of variance explained by its corresponding component, thereby providing a direct measure of that component's contribution to the overall data structure. The sum of these eigenvalues can be used to compute the cumulative percentage of variance explained, often referred to as the "quality of representation" (Harman, 1976). This cumulative measure offers a practical and intuitive criterion for assessing the degree to which a reduced set of components captures the intrinsic variability of the dataset, facilitating a balance between dimensionality reduction and information preservation (Demšar et al., 2013).

A widely adopted heuristic for determining the optimal number of components is the Kaiser criterion, which recommends retaining only those components whose eigenvalues exceed the value of one. This guideline is conceptually grounded in the fact that the total number of eigenvalues equals the number of active variables, yielding an average eigenvalue of one (Harman, 1976). Components with eigenvalues greater than this threshold, therefore, contribute more than the average variable to the total variance and are considered informative for data summarization. In the present study, we adhere to this criterion, systematically identifying components with eigenvalues above one, and calculating their cumulative contribution to the total variance, thereby ensuring that only the most informative dimensions were retained for subsequent analysis (Jolliffe, 2002).

To rigorously assess the appropriateness of PCA for the dataset under consideration, we employed two complementary diagnostic tests: Kaiser–Meyer–Olkin (KMO) and Bartlett's test of sphericity, testing whether the observed correlations are significantly different from zero (Harman, 1976). Together, these diagnostics provide robust evidence for the suitability of PCA as an analytical tool for capturing the underlying structure of the dataset. Beyond the assessment of global suitability, the contribution of individual variables and observations to each principal component was examined in detail through the computation of factor loadings and component coordinates. Factor loadings, or factor coordinates, quantify the influence of each original variable on the construction of individual components, offering nuanced insight into the relative importance and interpretive meaning of each variable within the transformed space (Benzécri, 1973).

Concurrently, the coordinates of each observation on the factor axes allow for an evaluation of the weight and influence of each case in shaping the variance captured by each component (Jombart et al., 2008). Observations with larger coordinates exert a greater leverage on the orientation and interpretation of the corresponding factor, revealing the extent to which individual cases drive the emergent structure identified by the PCA (Lloyd, 2010). Such a dual-level analysis, integrating both variables and observations, enhances the interpretability of the multivariate transformation and supports more informed conclusions regarding the patterns embedded within complex, high-dimensional datasets (Li et al., 2017).

5.2.4 Empirical application

A total of three PCA trials were conducted, each yielding components ordered according to the variance explained. Eigenvalues were computed, and the Kaiser criterion was applied to retain the most informative components. Factor loadings were interpreted to identify the variables contributing most to the observed patterns, while the scores of observations on the principal axes provided insights into spatial clustering and outlier detection (Lloyd, 2010). A comprehensive spatial matrix was constructed to characterize the landscape structure of metropolitan Athens, Greece, encompassing 115 administrative units, using high-resolution geospatial data obtained from the Urban Atlas (UA) initiative. The dataset, corresponding to the 2018 update of the UA program, provides standardized and comparable land-use maps at a 1:10,000 scale for over 300 major urban areas across the European Union. The Urban Atlas employs a hierarchical classification system encompassing 20 land-use categories, primarily derived from the Corine Land Cover nomenclature, which ensures thematic consistency and facilitates cross-city comparisons across Europe.

For the purposes of this study, the spatial extent of each land-use class was computed within the boundaries of each municipality in metropolitan Athens, Greece, following the methodology outlined in previous research. To quantify landscape composition at the municipal scale, we adopted the "zonal statistics" tool within the ArcGIS (ESRI, Redwoods, CA) software environment. This geo-processing function allows for the systematic extraction of statistical summaries – specifically, the proportion of each land-use class contained within a defined spatial unit – from raster (or vector-based) land-use datasets. This procedure ensured the generation of a consistent and reproducible dataset, capturing both the heterogeneity and the spatial distribution of land-use types across the metropolitan area.

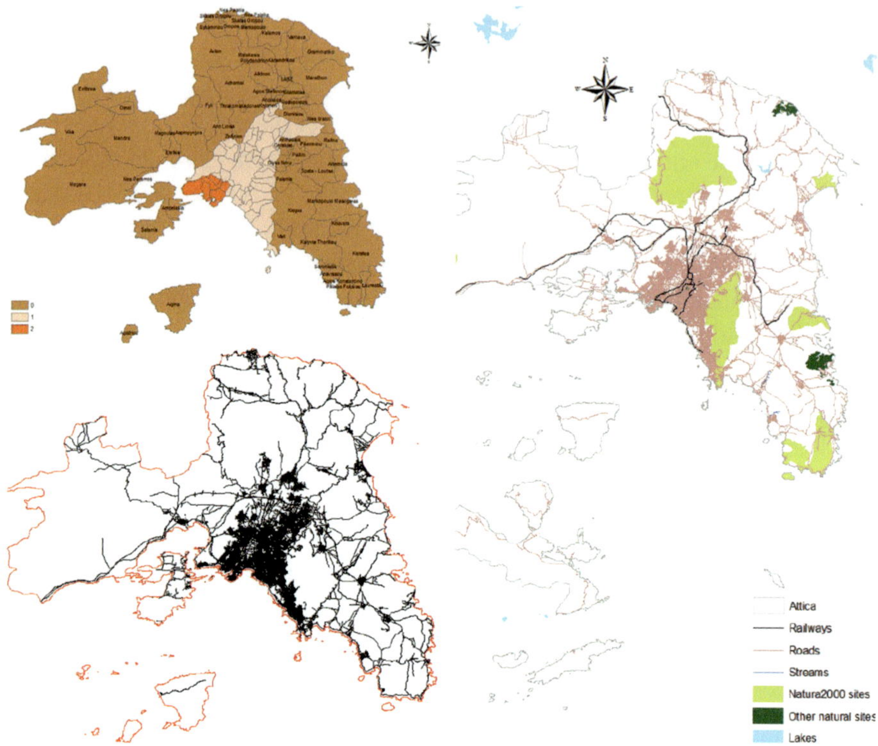

Figure 5.1 A visual description of the study area; upper left: an administrative map of local municipalities in metropolitan Athens; lower left: the road network; right: protected areas and infrastructures; source: authors' elaboration of official statistics and public digital maps.

To mitigate potential distortions arising from multi-collinearity among land-use variables, the class labeled "50,000" – typically representing undefined or non-urban categories – was excluded prior to subsequent analyses. The resulting data matrix comprised 19 columns corresponding to distinct land-use classes and 115 rows representing the local administrative units. This structured and standardized dataset provided a robust foundation for principal component analysis (PCA), enabling the investigation of spatial patterns, the quantification of land-use variability, and the identification of latent gradients in the spatial organization of metropolitan landscapes (Figure 5.1).

Beyond its immediate analytical utility, this approach facilitates a more nuanced understanding of how land-use composition reflects broader processes of urban expansion, functional differentiation, and socio-environmental interactions within metropolitan contexts. By systematically capturing the distribution of diverse land-use types, the matrix also allows for the integration of ecological, planning, and socioeconomic considerations, highlighting both the structural and functional dimensions of landscape. In this way, the dataset serves as a versatile tool for multivariate exploration and comparative analysis, providing insights into the spatial configuration and organizational principles that underpin contemporary urban systems.

5.3 Facts and Findings

Table 5.1 summarizes the diagnostic results of the principal component analysis (PCA) conducted using three distinct correlation matrices – Pearson, Spearman, and Kendall – considering only the principal components with eigenvalues greater than 1. As anticipated, the first principal component explained the largest proportion of total variance when the Pearson correlation matrix was employed, reflecting its sensitivity to linear relationships within the dataset. However, from the second component onward, the cumulative variance captured by PCA based on the Spearman correlation matrix consistently exceeded that obtained from both the Pearson and Kendall matrices. This pattern indicates that the non-parametric Spearman approach is more effective in detecting additional, potentially non-linear or monotonic structures within the data beyond the primary principal axis. The adequacy of the PCA for this dataset was further evaluated through two established diagnostic measures. Across the three correlation matrices, the PCA performed using Spearman coefficients achieved the highest scores for both diagnostics, highlighting its superior (overall) suitability for factor analysis.

The Pearson-based PCA demonstrated moderately strong performance, whereas the Kendall-based analysis exhibited comparatively lower diagnostic values, suggesting a reduced capacity to capture the underlying structure of the data. A detailed examination of the variance explained by each principal component across the three correlation metrics revealed that, although the Pearson matrix accounted for the largest proportion of variance in the first component, Spearman consistently outperformed the other matrices from the second component onward. This progressive gain in explained variance underscores the capacity of the Spearman-based PCA to uncover additional latent structures, particularly those reflecting monotonic or non-linear relationships.

Conversely, the Kendall correlation matrix proved less effective in enhancing component extraction within this dataset, resulting in a comparatively lower cumulative variance and suggesting a limited sensitivity to complex inter-variable patterns. Further examination of the component loadings corroborates these observations, reinforcing the conclusion that PCA based

Table 5.1 Principal component analysis outcomes by computation metric (KMO: Kayser–Meier–Olkin measure); source: authors' elaboration of official statistics.

	Percent variance explained by the component					KMO	Bartlett's
Coefficient	**1st**	**2nd**	**3rd**	**4th**	**5th**	**test**	**test**
Pearson	35.6	13.6	10.8	7.2	6.6	0.68	1712
Spearman	32.8	18.9	11.2	7.3	6.0	0.84	1771
Kendall	26.1	15.3	10.0	7.2	6.0	0.79	819

on Spearman coefficients provides a superior quality of representation, particularly for the first two principal components. Moreover, a comparison of case coordinates across the three PCA implementations reveals that, although all methods yielded broadly comparable spatial distributions, the Spearman-based PCA exhibited slightly enhanced discriminatory capacity, enabling more nuanced differentiation among cases along the primary axes.

5.4 Discussion

The present study employed a multidimensional framework to evaluate metropolitan landscape composition, adopting a generalized application of principal component analysis (PCA) that incorporates alternative correlation matrices (Demšar et al., 2013). The findings confirm that the selection of a correlation metric is not a trivial technical choice but one that fundamentally shapes the performance, interpretability, and robustness of PCA (Jolliffe & Cadima, 2016). In particular, the implementation based on Spearman's rank correlation consistently outperformed the variants constructed with Pearson and Kendall coefficients, both in terms of statistical diagnostics, as well as in the proportion of variance explained across successive components (Smarandache, 2009).

The superiority of the Spearman-based approach lies in its non-parametric nature, which renders it inherently less vulnerable to the presence of outliers and violations of normality – conditions that are especially pervasive in land-use and geospatial datasets characterized by heterogeneity, fragmentation, and non-linear relationships (Salvati & Zitti, 2005). By capturing monotonic trends, whether strictly linear or not, Spearman correlation enhances the capacity of PCA to reveal hidden structures and interdependencies that might otherwise remain obscured under traditional linear assumptions (Brouwer and Chadwick, 1991; Turner et al., 2007; Jombart et al., 2008).

While standard PCA identifies mathematically optimal directions of maximum variance reduction, minimizing squared error within a dataset, its reliance on linear correlation coefficients makes it prone to distortions (Coppi

& Bolasco, 1989). Conventional strategies often recommend the removal or transformation of such values. However, in large-scale and high-dimensional spatial datasets, such interventions may neither be feasible nor conceptually justified (Lloyd, 2010). Indeed, the very notion of an outlier can be ambiguous in spatial analysis, as what appears as an exception may instead be a critical expression of territorial specificity (Ferrara et al., 2016; Coluzzi et al., 2022; Seifollahi-Aghmiuni et al., 2022). In this light, the methodological advantage of Spearman correlation becomes evident: it offers an intrinsic form of robustness without necessitating preliminary data manipulation, thereby maintaining the integrity and diversity of the dataset (Conover, 1999).

Recent developments in robust, weighted, and non-linear extensions of PCA reflect a broader effort to adapt multivariate techniques to the complexities of contemporary data environments (Linting et al., 2007). Weighted PCA, for instance, introduces a hierarchy of importance across observations, which can be particularly relevant in spatial contexts where certain areas exert disproportionate influence due to demographic density or ecological value (Lamonica et al., 2020). Similarly, kernel-based and manifold learning generalizations attempt to model non-linear structures more explicitly, providing flexible representations of spatial heterogeneity (Vyas & Kumaranayake, 2006; Li et al., 2017; Montano & Jombart, 2017).

Yet even in the face of these advances, the pragmatic balance of interpretability, computational feasibility, and robustness achieved by Spearman correlation gives it a distinctive relevance for urban studies and landscape ecology (Turner et al., 2001; Wu & Hobbs, 2002; Li & Wu, 2004). It allows researchers to bridge the gap between mathematical abstraction and the messy, contingent realities of urban-rural territories, where spatial patterns emerge from the interplay of social, economic, and environmental forces rather than from idealized linear dynamics (Seto & Kaufmann, 2003; Seto & Fragkias, 2005; Egidi et al., 2020).

Moreover, the implications of adopting non-parametric correlation metrics extend beyond statistical refinement (Biasi et al., 2019). By offering a more faithful representation of the multifaceted character of complex landscapes (Francaviglia et al., 2019), such methods enrich the interpretive frameworks available to planners, geographers, and environmental analysts (Salvati et al., 2012; Tomao et al., 2017; Ferrara et al., 2020).

Decisions on land-use allocation, ecological conservation, or infrastructural development increasingly rely on tools capable of revealing latent dimensions of spatial organization that are not reducible to simple linear correlations (Salvati & Carlucci, 2011; Colantoni et al., 2015; Salvati et al., 2017). The enhanced discrimination achieved through Spearman-based PCA thus not only strengthens methodological rigor but also broadens the epistemological horizon of regional science, enabling a more nuanced dialogue between quantitative analysis and the interpretive traditions of landscape studies (Salvati & Zitti, 2008; Salvati et al., 2013; Salvati & Carlucci, 2014).

As the availability of high-resolution spatial data continues to expand and the demand for quantitative, cross-disciplinary approaches grows – particularly in applications at fine spatial scales – there emerges an increasingly urgent need to refine and advance multivariate analytical methods capable of coping with such complexity (Turner et al., 2007; Khatun, 2009; Kosmas et al., 2015). Conventional PCA has undoubtedly proven useful in exploratory analysis, yet its reliance on assumptions of linearity and sensitivity to outliers often constrains its interpretive power when applied to heterogeneous and multidimensional datasets (Sokal & Rohlf, 2012).

In this context, the integration of non-parametric correlation metrics, such as Spearman's rank coefficients, becomes not only a methodological improvement but also a conceptual advancement, as it enables the identification of monotonic and potentially non-linear relationships that are frequently hidden in ecological data structures (Hollander & Wolfe, 1999). By broadening the analytical lens, researchers can capture more faithfully the subtle dynamics that characterize settlement morphologies, land-use mosaics, and environmental gradients (Mancino et al., 2016).

The implications of such methodological refinements are profound (Salvati et al., 2013). In spatial planning, for instance, robust non-parametric PCA can support the detection of latent structures that govern the spatial organization of land-use functions, revealing clusters, contrasts, or anomalies that would otherwise remain obscured (Salvati, 2013; Vardopoulos et al., 2021; Seifollahi-Aghmiuni et al., 2022). In environmental monitoring, the capacity to handle outliers without distortion allows for a more accurate assessment of ecological thresholds, pollution patterns, or landscape fragmentation (Ferrara et al., 2017; Francaviglia et al., 2019; Coluzzi et al., 2022). Similarly, in landscape ecology, these approaches foster a deeper comprehension of the interplay between natural systems and human modifications, shedding light on emergent spatial patterns that cannot be fully apprehended through traditional parametric tools (Turner et al., 2001; Wu & Hobbs, 2002; Li & Wu, 2004).

Future research should therefore continue to evaluate and compare alternative generalizations of PCA, extending beyond non-parametric correlations to explore non-linear, weighted, and robust formulations (Jolliffe & Cadima, 2016). Each of these variants has the potential to improve the quality of representation by emphasizing different structural features of the data, whether it be the preservation of local neighborhoods, the reduction of noise, or the differential weighting of variables according to ecological or social significance (Vyas & Kumaranayake, 2006; Montano & Jombart, 2017; Lamonica et al., 2020). The challenge lies in tailoring methodological innovation to the epistemological needs of spatial disciplines, ensuring that statistical sophistication translates into meaningful insights for both theory and practice (Lambin & Meyfroidt, 2011).

Such methodological progress will be instrumental in enhancing decision-making processes in contexts where scale, complexity, and uncertainty converge (e.g., Quaranta et al., 2020). From designing resilient and aesthetically coherent urban forms, to safeguarding ecological integrity, to developing adaptive planning strategies in the face of climate change, the capacity to extract robust and nuanced signals from large-scale spatial datasets represents a decisive step forward (Salvati et al., 2015; Tomao et al., 2017; Biasi et al., 2019). Ultimately, advancing PCA through non-parametric and generalized approaches does not merely refine the analytical toolkit: it enriches our ability to interpret, represent, and shape the evolving landscapes in which contemporary societies are embedded (Salvati & Zitti, 2009; Salvati et al., 2012; Ferrara et al., 2016, 2020).

5.5 Concluding Remarks

The implications of these findings extend beyond statistical refinement. Metropolitan landscapes are the product of multiple interacting processes – economic, environmental, social, and morphological – that often generate non-linear patterns of association among relevant variables (or indicators) and complex (latent) data structures that are eminently difficult to analyze and ascertain in their articulation. For instance, the relationship between residential density and commercial land-use may not evolve linearly but instead follow threshold effects, where sudden increases in density trigger a disproportionate rise in commercial activity. Traditional linear metrics risk obscuring such dynamics, whereas rank-based correlations are better suited to detect these relationships. Such distinctions, however, are crucial in metropolitan regions where historical layering, rapid settlement expansion, and uneven socioeconomic development have produced highly complex spatial mosaics.

By demonstrating the added value of non-parametric correlation measures, the study suggests that PCA can be re-conceptualized as a more flexible and context-sensitive tool for urban studies and regional science. This methodological advance holds significance for a broad spectrum of applications.

In spatial planning, more robust PCA models can provide clearer insights into functional diversity, spatial segregation, and land-use complementarities, supporting evidence-based strategies for sustainable development. In environmental assessment, they can definitely improve the detection of ecological gradients and the identification of areas particularly vulnerable to urbanization pressures. In landscape ecology, they can facilitate the analysis of spatial heterogeneity and connectivity, thus informing policies aimed at conserving biodiversity within metropolitan regions.

References

Benzécri, J.-P. (1973). L'analyse des données. Paris: Dunod.

Biasi, R., Brunori, E., Ferrara, C., & Salvati, L. (2019). Assessing impacts of climate change on phenology and quality traits of Vitis vinifera L.: The contribution of local knowledge. Plants, 8(5), 121.

Brouwer, F. B., Thomas, A. J., & Chadwick, M. J. (1991). Land use changes in Europe: Processes of change, environmental transformations and future patterns. Dordrecht: Kluwer.

Colantoni, A., Ferrara, C., Perini, L., & Salvati, L. (2015). Assessing trends in climate aridity and vulnerability to soil degradation in Italy. Ecological Indicators, 48, 599–604.

Coluzzi, R., Bianchini, L., Egidi, G., Cudlin, P., Imbrenda, V., Salvati, L., & Lanfredi, M. (2022). Density matters? Settlement expansion and land degradation in peri-urban and rural districts of Italy. Environmental Impact Assessment Review, 92, 106703.

Conover, W. J. (1999). Practical nonparametric statistics (3rd ed.). Wiley.

Coppi, R., & Bolasco, S. (1989). Multiway data analysis. Amsterdam: North Holland.

Demšar, U., Harris, P., Brunsdon, C., Fotheringham, A. S., & McLoone, S. (2013). Principal component analysis on spatial data: An overview. Annals of the Association of American Geographers, 103(1), 106–128.

Egidi, G., Salvati, L., & Vinci, S. (2020). The long way to Tipperary: City size and worldwide urban population trends, 1950–2030. Sustainable Cities and Society, 60, 102148.

Ferrara, A., Kelly, C., Wilson, G. A., Nolè, A., Mancino, G., Bajocco, S., & Salvati, L. (2016). Shaping the role of 'fast' and 'slow' drivers of change in forest-shrubland socio-ecological systems. Journal of Environmental Management, 169, 155–166.

Ferrara, A., Kosmas, C., Salvati, L., Padula, A., Mancino, G., & Nolè, A. (2020). Updating the MEDALUS ESA framework for worldwide land degradation and desertification assessment. Land Degradation & Development, 31(12), 1593–1607.

Ferrara, C., Carlucci, M., Grigoriadis, E., Corona, P., & Salvati, L. (2017). A comprehensive insight into the geography of forest cover in Italy: Exploring the importance of socioeconomic local contexts. Forest Policy and Economics, 75, 12–22.

Francaviglia, R., Di Bene, C., Farina, R., Salvati, L., & Vicente-Vicente, J. L. (2019). Assessing "4 per 1000" soil organic carbon storage rates under Mediterranean climate: A comprehensive data analysis. Mitigation and Adaptation Strategies for Global Change, 24, 795–818.

Harman, H. H. (1976). Modern factor analysis (3rd ed.). Chicago: University of Chicago Press.

Hollander, M., & Wolfe, D. A. (1999). Nonparametric statistical methods (2nd ed.). Wiley, London.

Incerti, G., Feoli, E., Salvati, L., Brunetti, A., & Giovacchini, A. (2007). Analysis of bioclimatic time series and their neural network-based classification to characterise drought risk patterns in South Italy. International Journal of Biometeorology, 51, 253–263.

Johnson, R. A., & Wichern, D. W. (2002). Applied multivariate statistical analysis (5th ed.). Prentice Hall, New York.

Jolliffe, I. T., & Cadima, J. (2016). Principal component analysis: A review and recent developments. Philosophical Transactions of the Royal Society A: Mathematical, Physical and Engineering Sciences, 374(2065), 20150202.

Jolliffe, I. T. (2002). Principal component analysis (2nd ed.). New York: Springer.

Jombart, T., Devillard, S., Dufour, A. B., & Pontier, D. (2008). Revealing cryptic spatial patterns in genetic variability by a new multivariate method. Heredity, 101(1), 92–103.

Kendall, M. G. (1970). Rank correlation methods (4th ed.). London: Charles Griffin & Company.

Khatun, T. (2009). Measuring environmental degradation by using principal component analysis. Environment, Development and Sustainability, 11, 439–457.

Kosmas, C., Detsis, V., Karamesouti, M., Kounalaki, K., Vassiliou, P., & Salvati, L. (2015). Exploring long-term impact of grazing management on land degradation in the socio-ecological system of Asteroussia Mountains, Greece. Land, 4(3), 541–559.

Lambin, E. F., & Meyfroidt, P. (2011). Global land use change, economic globalization, and the looming land scarcity. Proceedings of the National Academy of Sciences, 108(9), 3465–3472.

Lamonica, G. R., Recchioni, M. C., Chelli, F. M., & Salvati, L. (2020). The efficiency of the cross-entropy method when estimating the technical coefficients of input–output tables. Spatial Economic Analysis, 15(1), 62–91.

Lavit, C., Escoufier, Y., Sabatier, R., & Traissac, P. (1994). The ACT (STATIS) method. Computational Statistics & Data Analysis, 18, 97–119.

Li, H., & Wu, J. (2004). Use and misuse of landscape indices. Landscape Ecology, 19, 389–399.

Li, J., Reich, B. J., & Bondell, H. D. (2017). Extending principal component analysis for spatially correlated data. BMC Bioinformatics, 18, 1–13.

Linting, M., Meulman, J. J., Groenen, P. J. F., & van der Kooij, A. (2007). Nonlinear principal components analysis: Introduction and application. Journal of Statistical Software, 12(3), 1–58.

Lloyd, C. D. (2010). Analysing population characteristics using geographically weighted principal components analysis: A case study of Northern Ireland in 2001. Computers, Environment and Urban Systems, 34, 389–399.

Mancino, G., Nolè, A., Salvati, L., & Ferrara, A. (2016). In-between forest expansion and cropland decline: A revised USLE model for soil erosion risk under land-use change in a Mediterranean region. Ecological Indicators, 71, 544–550.

Meyfroidt, P., & Lambin, E. F. (2008). The causes of the reforestation in Vietnam. Land Use Policy, 25(2), 182–197.

Montano, V., & Jombart, T. (2017). An eigenvalue test for spatial principal component analysis. BMC Bioinformatics, 18, 1–13.

Quaranta, G., Salvia, R., Salvati, L., Paola, V. D., Coluzzi, R., Imbrenda, V., & Simoniello, T. (2020). Long term impacts of grazing management on land degradation in a rural community of Southern Italy: Depopulation matters. Land Degradation & Development, 31(16), 2379–2394.

Sallustio, L., Pettenella, D., Merlini, P., Romano, R., Salvati, L., Marchetti, M., & Corona, P. (2018). Assessing the economic marginality of agricultural lands in Italy to support land use planning. Land Use Policy, 76, 526–534.

Salvati, L. (2013). Monitoring high-quality soil consumption driven by urban pressure in a growing city (Rome, Italy). Cities, 31, 349–356.

Salvati, L., & Carlucci, M. (2011). The economic and environmental performances of rural districts in Italy: Are competitiveness and sustainability compatible targets? Ecological Economics, 70(12), 2446–2453.

Salvati, L., & Carlucci, M. (2014). A composite index of sustainable development at the local scale: Italy as a case study. Ecological Indicators, 43, 162–171.

Salvati, L., & Zitti, M. (2005). Land degradation in the Mediterranean basin: Linking bio-physical and economic factors into an ecological perspective. Biota, 6, 67–77.

Salvati, L., & Zitti, M. (2008). Regional convergence of environmental variables: Empirical evidences from land degradation. Ecological Economics, 68(1–2), 162–168.

Salvati, L., & Zitti, M. (2009). Substitutability and weighting of ecological and economic indicators: Exploring the importance of various components of a synthetic index. Ecological Economics, 68(4), 1093–1099.

Salvati, L., Mavrakis, A., Colantoni, A., Mancino, G., & Ferrara, A. (2015). Complex adaptive systems, soil degradation and land sensitivity to desertification: A multivariate assessment of Italian agro-forest landscape. Science of the Total Environment, 521, 235–245.

Salvati, L., Morelli, V. G., Rontos, K., & Sabbi, A. (2013). Latent exurban development: City expansion along the rural-to-urban gradient in growing and declining regions of southern Europe. Urban Geography, 34(3), 376–394.

Salvati, L., Perini, L., Sabbi, A., & Bajocco, S. (2012). Climate aridity and land use changes: A regional-scale analysis. Geographical Research, 50(2), 193–203.

Salvati, L., Ranalli, F., Carlucci, M., Ippolito, A., Ferrara, A., & Corona, P. (2017). Forest and the city: A multivariate analysis of peri-urban forest land cover patterns in 283 European metropolitan areas. Ecological Indicators, 73, 369–377.

Salvati, L., Venanzoni, G., Serra, P., & Carlucci, M. (2016). Scattered or polycentric? Untangling urban growth in three southern European metropolitan regions through exploratory spatial data analysis. The Annals of Regional Science, 57, 1–29.

Salvati, L., Zitti, M. (2009). The environmental 'risky' region: Identifying land degradation processes through integration of socio-economic and ecological indicators in a multivariate regionalization model. Environmental Management, 44(5), 888–899.

Seifollahi-Aghmiuni, S., Kalantari, Z., Egidi, G., Gaburova, L., & Salvati, L. (2022). Urbanisation-driven land degradation and socioeconomic challenges in peri-urban areas: Insights from Southern Europe. Ambio, 51(6), 1446–1458.

Seto, K. C., & Fragkias, M. (2005). Quantifying spatiotemporal patterns of urban land-use change in fourteen metropolitan areas of China. Land Use Policy, 22(3), 144–156.

Seto, K. C., & Kaufmann, R. K. (2003). Modeling the drivers of urban land use change in the Pearl River Delta, China: Integrating remote sensing with socioeconomic data. Land Economics, 79(1), 106–121.

Smarandache, F. (2009). Alternatives to Pearson's and Spearman's correlation coefficients. International Journal of Statistics and Economics, 3(S09), 47–53.

Sokal, R. R., & Rohlf, F. J. (2012). Biometry: The principles and practice of statistics in biological research (4th ed.). W. H. Freeman and Company.

Spearman, C. (1904). The proof and measurement of association between two things. The American Journal of Psychology, 15(1), 72–101.

Srivastava, P. K., Mukherjee, S., & Gupta, M. (2010). Impact of urbanization on land use/land cover change using remote sensing and GIS: A case study. International Journal of Ecological Economics and Statistics, 18(S10), 106–117.

Tomao, A., Quatrini, V., Corona, P., Ferrara, A., Lafortezza, R., & Salvati, L. (2017). Resilient landscapes in Mediterranean urban areas: Understanding factors influencing forest trends. Environmental Research, 156, 1–9.

Turner, B. L., Lambin, E. F., & Reenberg, A. (2007). The emergence of land change science for global environmental change research. Global Environmental Change, 17(4), 211–226.

Turner, M. G., Gardner, R. H., & O'Neill, R. V. (2001). Landscape ecology in theory and practice. Berlin: Springer.

Vardopoulos, I., Tsilika, E., Sarantakou, E., Zorpas, A. A., Salvati, L., & Tsartas, P. (2021). An integrated SWOT-PESTLE-AHP model assessing sustainability in adaptive reuse projects. Applied Sciences, 11(15), 7134.

Vyas, S., & Kumaranayake, L. (2006). Constructing socio-economic status indices: How to use principal components analysis. Health Policy and Planning, 21(6), 459–468.

Wu, J., & Hobbs, R. J. (2002). Key issues and research priorities in landscape ecology: An idiosyncratic synthesis. Landscape Ecology, 17, 355–365.

6

"Burnt landscapes": Improving Wildfire Management in Sensitive Hotspots

Alessia D'Agata[1,2], Marco Maialetti[3], Donato Scarpitta[4], Vito Imbrenda[5], Pierpaolo Lombardi[6], and Luca Salvati[1]

[1]Department of Methods and Models for Economics, Territory and Finance (MEMOTEF), Faculty of Economics, Sapienza University of Rome, Italy
[2]Department of Geography, Building B, Autonomous University of Barcelona, Spain
[3]Independent Researcher and Lawyer, Italy
[4]Independent Researcher and Architect, Italy
[5]Institute of Methodologies for Environmental Analysis, Italian National Research Council (CNR-IMAA), Italy
[6]Gruppo Maurizi, Italy

Email: alessia.dagata@uniroma1.it; avv.marcomaialetti@gmail.com; scarpitta.donato@libero.it; vito.imbrenda@cnr.it; pierpaolo.lombardi@gmail.com; luca.salvati@uniroma1.it

Abstract

Fire prevention strategies have to deal with evolving environmental and socioeconomic scenarios. The combination of a warming climate, fuel patterns, and socioeconomic conditions is responsible for increasing large fires throughout Europe. In the last decades, simultaneous large fires often overwhelmed the suppression capacity of single countries – not only in fire-prone areas, but also in countries occasionally affected by such events. In the latter countries, wildfires result in more severe and long-lasting effects due to the lower fire resistance and resilience of forests. In order to cope with a changing world, the transfer of fire management strategies is mandatory, and original proposals on prevention actions based on broader sustainable land/forest management may increase wildfire response preparedness. These objectives

DOI: 10.1201/9788743811107-6

will be reached by sharing national, regional, and local good practices among countries with different operational experiences and skills, underlying different levels of response preparedness.

6.1 Introduction

Despite a slightly decreasing trend of yearly-burned area, large wildfires (i.e., fires larger than 100 hectares) are rising in frequency and severity in many European countries (Birot, 2009; Pausas et al., 2009; Moreira et al., 2011), not only in traditionally fire-prone countries, with increasing environmental and socioeconomic costs (Abatzoglou et al., 2021; Akter & Grafton, 2021; Wunder et al., 2021). These large fires negatively affect wide-scale ecosystems, human infrastructure, and frequently result in unprecedented fatalities, with fire outcomes being more evident and dangerous where urbanization is eroding natural areas and land resources (Seifollahi-Aghmiuni et al., 2022). Fire risk is also triggered by land abandonment in many parts of Europe (e.g., Viedma et al., 2015). As a matter of fact, changes in socioeconomic conditions and political decisions resulted in the depopulation of rural areas, thus reducing protection of forests from natural threats, including human-caused wildfires (Ferrara et al., 2016; Mancino et al., 2016; Sallustio et al., 2018). In addition, climate change scenarios predict drier seasons and warmer climates, increasing water deficits and susceptibility of forest systems to natural disasters (Vitolo et al., 2019, 2020).

Since the early 2000s, such changes have been producing a shift toward the north in severe and intense wildfires, previously observed mostly in Mediterranean Europe (Viedma et al., 2018). Species admixtures and growth trends of forests are very likely to be modified even in continental areas (Nocentini et al., 2017), where wildfires are currently considered just an unlikely event. In this regard, while more drastic conditions are expected in fire-prone areas, new challenges may arise at northern latitudes, where fire severity may be significantly higher and long-lasting than in fire-prone, adapted ecosystems (Ascoli et al., 2021). With this view in mind, sustainable land management policies oriented toward an effective fire management governance are asked to incorporate an improved knowledge of spatially heterogeneous wildfire drivers (Fernandes et al., 2011; Salvati & Carlucci, 2011; Tomao et al., 2017). An improved planning for wildfire prevention requires considering a large ensemble of (changing over time) environmental and socioeconomic conditions at the continental scale in Europe (Corona et al., 2015; Tedim et al., 2016; Verkerk et al., 2018).

Research has outlined the importance of fire prevention measures in effective wildfire governance, especially to improve wildfire response preparedness. However, the perceptions about fire management adaptive measures show contrasting approaches between managers and researchers (Raftoyannis et al., 2014), thus highlighting the need to strengthen the collaboration between "scientists" and "professionals." An improved governance of fire prevention based on relevant experiences and good practices of sustainable land/forest management has been demonstrated to reduce forest vulnerability to wildfires (Ascoli et al., 2013; Santopuoli et al., 2015; Ferrara et al., 2017). However, an improved understanding of the applicability and effectiveness of wildfire prevention actions by linking sustainable land management with fire regimes over different territorial conditions is increasingly required in a global change context (Fares et al., 2017). In particular, the identification of actions specifically targeted to fire prevention and their integration within broader measures of sustainable land and forest management is considered a fundamental strategy to improve wildfire response preparedness (Colantoni et al., 2015). This deserving issue may benefit from extensive research networks supporting (or developing ex-novo) methodologies for a comprehensive and comparative assessment of the "wildfire issue" within a renewed "prevention" vision fully integrated in broader strategies of sustainable land management (Incerti et al., 2007; Salvati et al., 2012; Kosmas et al., 2015). In this regard, an increasing institutional awareness of the importance of sustainable land/management is seen as an effective factor boosting preparedness to large fire management at various governance scales (Salvati & Carlucci, 2014).

Research dissemination-based networks are increasingly required to (i) identify good practices in fire prevention, (ii) boost management actions at different governance levels, and (iii) improve human resources/skills for developing high-quality fire management governance, improving trans-national cooperation between most-prone and less-prone countries, and contributing to protecting environments and biodiversity (Figure 6.1). An enforced trans-national scientific cooperation sharing relevant experiences in fire prevention and coordination improvement among the relevant institutions involved in sustainable forest management represents the necessary actions to fill this knowledge gap (Salvati & Zitti, 2008; Ferretti et al., 2014; Wunder et al., 2021). Based on the inherent complexity of the "wildfire" issue, scientific networking will give an added value to research outcomes, advancing cooperative knowledge and practical solutions to multi-dimensional problems (Zuur et al., 2010).

Figure 6.1 Biodiversity at risk in fire-prone ecosystems; source: authors' photographic archive.

6.2 Conceptual Background

Although still often neglected, fire management should be a major issue in forest policy. Wildfires affect nearly 0.25 million hectares every year in Europe, representing one of the prominent hazards that could turn into disasters (Incerti et al., 2007). Fire management governance is currently experiencing an evolving fire scenario due to the combination of climate change, fuel patterns, land-use conflicts, and socioeconomic conditions (Salvati & Zitti, 2008; Colantoni et al., 2015; Ferrara et al., 2016). These cause concurrent large fires that might overwhelm the suppression capacity of any single country. Furthermore, large wildfires are discontinuous over time and space because of increased fluctuations of the main drivers, thus lowering the attention of policy and media (Moreira et al., 2011).

In recent times, a shift of large fires' distribution toward the north was observed in Europe: relevant fire events, usually recorded in Southern Europe, and to a lesser extent, in some countries of Western and Eastern Europe, increasingly occur in less prone countries, such as in Northern

Europe (Viedma et al., 2015, 2018). A crucial issue is thus to increase advanced research knowledge and policy stakeholders' awareness about major risk factors that influence large fires to occur in each region and identify which changes should be introduced in the current wildfire management governance to cope with such extreme fire seasons (Raftoyannis et al., 2014). In this regard, formal and informal experiences in fire prevention at different spatial and governance scales in the most prone European countries form a relevant core of effective actions to be shared cross-country and adapted to different environmental and socioeconomic contexts characterized by low (and increasing) fire risk and less experience in large wildfire prevention. Advanced know-how in such fields will allow for an effective alignment of wildfire research and policy based on a convergence toward scientific knowledge, operational approaches, and good practices across European countries.

Boosting scientific cooperation on fire prevention is particularly relevant in the present socioeconomic context. Since limited public funding for fire prevention – which requires significant monetary resources and operational knowledge – is available throughout Europe, a research-based network sharing relevant experiences and best practices in sustainable land/forest management supporting fire prevention is particularly appropriate in both fire-prone countries (Southern, and to a lesser extent, Western and Eastern Europe) and less-prone contexts with a demonstrated increase in the level of vulnerability to wildfires (Moreira et al., 2011; Viedma et al., 2015; Vitolo et al., 2019).

To drive response preparedness, the development of appropriate fuel management strategies is becoming increasingly crucial for fire prevention in the context of global warming and increasing climate variability (Tedim et al., 2016). Implementing and applying dedicated fire prevention and management abilities to differential fire regimes at the regional and local scale would lead to the reduction of wildfire occurrence and mitigation of wildfire severity, containing operational costs, with indirect benefits for public budgets (Akter & Grafton, 2021).

In this regard, scientific cooperation focusing on the identification and sharing of fire prevention finally allows for to integrate a multi-disciplinary knowledge linking good practices at the local scale with successful regional experiences in firefighting programs (Ascoli et al., 2013). To enhance a more effective fire-fighting capability, key players responsible for fire prevention at different spatial and governance levels have to coordinate actions, improving the level of response preparedness to large fires. An example of these needs is represented by a recently established network, namely the European Forest Fire Fighting Council (EuForFire), which aims at increasing cross-country cooperation and collaboration in firefighting training (Michetti & Pinar, 2019).

It includes six members from Portugal, France, Spain, and Italy, and represents shared visions on wildfire management mainly derived from Mediterranean experiences (Tomao et al., 2017). In order to define an efficient trans-national, pan-European strategy of fire prevention, an improved research network should integrate previously established cooperative experiences such as the EuForFire initiative (Vitolo et al., 2019). Active networking will delineate the positive impact of sustainable land management on institutional coordination at multiple spatial and governance scales (Ferretti et al., 2014).

6.2.1 The role of networking

Networking is becoming more and more important in wildfire management strategies at the European level. Several projects based on (or including) continuous exchange of experiences were carried out and finalized over the last two decades (Fernandes et al., 2011). However, most of these projects have preferentially addressed the multifaceted issue of fire suppression with a specific focus on the European ecological and socioeconomic contexts most affected by large fires. Additionally, international research projects such as FIRE PARADOX, FUME, and MEDWILDFIRELAB, among the others, have contributed significantly to the recent debate on wildfire dynamics addressing the complex interplay between fire prevention and suppression, and identifying critical aspects that influence the effectiveness of an integrated strategy boosting response preparedness in fire-prone countries (Fernandes et al., 2011; Fares et al., 2017; Wunder et al., 2021).

Taken together, earlier projects highlighted the increasing impact of specific fire drivers in fragile ecological systems and complex landscapes (Santopuoli et al., 2015; Nocentini et al., 2017; Viedma et al., 2018). In particular, operational and practical experiences deriving from earlier projects suggest that an increased cooperation and networking strategy boosting a common and shared ensemble of prevention actions seems to be a reasonable and efficient tool to enhance effective responses to large fires (Ascoli et al., 2021). It should be mentioned that some operational networks already exist in Europe, such as forest fire fighting training networks (e.g., EuForFire) or research networks (e.g., EUFIRELAB). However, a project capable of further increasing the connection between science and professionals, and translating knowledge into governance actions, is still missing.

The above-mentioned research initiatives call for further scientific cooperation, integrating all the aspects related to wildfire prevention strategies. The established networking activities will support a coherent analysis of research-based solutions for fire prevention, raising awareness of policy-makers to the importance of a novel governance of large wildfires (Vitolo et al., 2019). Advanced knowledge should be disseminated through an appropriate communication strategy with various research actors, contributing to the adoption of efficient technologies and good practices at the EU level. With these actions, the research findings and cooperation established in recent EU-funded projects should be consolidated, additionally considering more recent networking actions (Wunder et al., 2021).

6.2.2 Policy implementation

By reviewing and sharing long-term knowledge on good practices and evidencing how specific actions of fire prevention can reduce the frequency of large fires and make their suppression more cost-effective, an intense program of scientific cooperation within research institutions across Europe seems to be particularly appropriate (e.g., Salvati et al., 2012). This cooperative program is aimed at increasing advanced theoretical and practical knowledge on large wildfires' prevention, supporting a joint research agenda with practical solutions boosting response preparedness to large wildfires and an improved, knowledge-based governance of "fire prevention" oriented toward practical cooperation between institution members and policy stakeholders at both national and European levels (Wunder et al., 2021). These objectives will be reached by sharing national, regional, and local good practices among countries with different operational experiences and skills, underlying different levels of wildfire response preparedness.

Since these actions should address wildfire issues intimately associated with sustainable forest/land-use, the "governance" topic should be prioritized, focusing on the design of appropriate "management and control" schemes through identification of relevant experiences at various governance scales and delivery of guidelines for optimizing forest fire prevention (Salvati & Carlucci, 2011). Based on novel approaches to involve stakeholders and encompassing all forest conditions of Europe, the identification of good practices will contribute to fire preparedness actions and governance, addressing regional diversity within Europe and the global issue of increasing large wildfires in less fire-prone countries, such as Northern regions (Seifollahi-Aghmiuni et al., 2022).

The main objectives of such strategies may include a pan-European common knowledge of the "wildfire issue" based on an improved understanding of prevention and preparedness actions (Santopuoli et al., 2015).

Boosting a trans-national research-dissemination network for coordinating European actions addressing the issue of large wildfires prevention and preparedness seems to be an additional topic of relevance in this direction (Wunder et al., 2021). In this perspective, improving a scientifically based awareness of prevention and preparedness actions among policymakers is a demanding task (Ferrara et al., 2017). Collaboration with (and support to) the existing organizations and networking initiatives is a further key point when developing such targets (Ferretti et al., 2014; Corona et al., 2015; Tomao et al., 2017). To achieve the above-mentioned objectives, joint research in the field of sustainable forest and land management should be stimulated, allowing identification of good governance, sharing scientific information, and fostering new technical knowledge (Salvati & Carlucci, 2014).

More specifically, transnational scientific cooperation will provide a common knowledge supporting renewed theoretical approaches and practical methodologies to identify and share relevant experiences and good practices among European countries along different geographical gradients in Europe, from Northern to Southern countries and from Western to Eastern countries (Viedma et al., 2015, 2018). These gradients reflect different wildfires' regimes and sensitivity to large and severe events, reflecting a spatially varied ensemble of ecological conditions, forest characteristics, socioeconomic, institutional, and policy contexts (Ascoli et al., 2021). Sharing research experiences and disseminating good practices across different environmental and socioeconomic conditions may clearly strengthen scientific cooperation between fire-prone and less affected countries (Santopuoli et al., 2015).

6.2.3 Decision-making and spatial planning

This approach is also expected to contribute to enforced scientific and technical knowledge, fueling policy awareness of the key role of fire prevention at the European level, with the final goal to unravel and best exploit research findings and maximize empirical outcomes from these results (Moreira et al., 2011; Fares et al., 2017; Vitolo et al., 2019). By sharing and spreading "smart" solutions locally implemented through transnational cooperation among project partners, and other research institutes, the proposed networking should finally develop a cross-country dissemination strategy improving a knowledge-based awareness of policy makers in the role of a common "wildfire prevention" governance, boosting response preparedness to large and severe events (Akter & Grafton, 2021).

Distinctively, the networking activities should be oriented toward the achievement of specific actions supporting harmonized research approaches to the collection of measurable (quali-quantitative) information on relevant experiences and good practices in sustainable forest/land management associated with fire prevention, providing a set of key data collected at local, regional, and country level across Europe. Achievable results derived from scientific cooperation will be grounded on a standardized literature review and database compilation, using quali-quantitative approaches, including metadata analysis (Santopuoli et al., 2015; Tedim et al., 2016; Michetti & Pinar, 2019). The final output of the research network will be a relevant set of common experiences, field measures, and policy actions in sustainable land management and fire prevention, translated into appropriate guidelines shared among institutions and disseminated to policy stakeholders. In this perspective, scientific networking definitely maximizes the added value of theoretical and applied research, providing a timely and significant contribution to knowledge building and policy implementation (Ascoli et al., 2021).

Generally speaking, promoting an integrated research network with specific competencies and expertise in the field of large fire prevention is the direct objective of a specific research cooperation in environmental and ecological fields (Corona et al., 2015). Prevention by fuel management measures, reduction of land-use conflicts, contrast of land abandonment, and urban containment strategies will allow evaluating functional solutions under vastly different background conditions, providing a cost-benefit assessment and sharing practices and policies in national and international contexts, even in the light of cross-border cooperation (Moreira et al., 2011; Tomao et al., 2017; Akter & Grafton, 2021).

Developing a set of quali-quantitative criteria to identify country-, regional-, and local-scale good strategies may support an effective production of techniques, materials, and approaches to raise awareness of policymakers (Fares et al., 2017). This active international network to spread research-based "smart solutions" to a wide range of stakeholders should finally increase awareness of the importance of the prevention-preparedness chain in reducing large fire risk (Seifollahi-Aghmiuni et al., 2022). Research cooperation will definitely consolidate the necessary expertise to translate the identified good practices into specific measures adapted to vastly different ecological and socioeconomic contexts representative of the inherent diversity in European forests.

6.3 Concluding Remarks

By identifying, reviewing, and summarizing good practices and operational errors and planning shortcomings in recent times, the envisioned strategy reveals novel insights to both scientific and technical literature. A clear view of the solutions for fire prevention will be produced, raising awareness of decision-makers on the application of integrated fire management strategies aimed at preventing the consequences of disastrous forest fires. These results – suitably documented with an appropriate communication strategy and training material – will contribute to the adoption of common good practices, for instance, at the European level. In this perspective, operational measures such as setting partnerships between different countries' team levels will be the basis for cross-boundaries collaboration and further cooperation in large wildfires management.

A knowledge exchange platform on large wildfire prevention may function as an active hub for expertise, experiences, and the sharing of good practices among research/institutional stakeholders. Networking results will foster new scientific knowledge and tools, with a positive side impact on country and continental wildfire governance, improving the effectiveness of prevention measures and boosting response preparedness, with an indirect reduction of the socioeconomic and environmental costs of suppression and post-fire land management actions.

References

Abatzoglou, J. T., Juang, C. S., Williams, A. P., Kolden, C. A., & Westerling, A. L. (2021). Increasing synchronous fire danger in forests of the western United States. Geophysical Research Letters, 48(2), e2020GL090844.

Acácio, V., Holmgren, M., Rego, F., Moreira, F., & Mohren, G. M. J. (2009). Are drought and wildfires turning Mediterranean cork oak forests into persistent shrublands? Agroforestry Systems, 76(2), 389–400.

Adler, N., & Yazhemsky, E. (2010). Improving discrimination in data envelopment analysis: PCA–DEA or variable reduction. European Journal of Operational Research, 202(1), 273–284.

Akter, S., & Grafton, R. Q. (2021). Do fires discriminate? Socio-economic disadvantage, wildfire hazard exposure and the Australian 2019–20 'Black Summer' fires. Climatic Change, 165(53).

Ascoli, D., Castagneri, D., Valsecchi, C., Conedera, M., & Bovio, G. (2013). Post-fire restoration of beech stands in the Southern Alps by natural regeneration. Ecological Engineering, 54, 210–217.

Ascoli, D., Moris, J. V., Marchetti, M., & Sallustio, L. (2021). Land use change towards forests and wooded land correlates with large and frequent wildfires in Italy. Annals of Silvicultural Research, 46(2), 177–188.

Birot, Y. (2009). Living with wildfires: What science can tell us – A contribution to the science-policy dialogue (No. 15). European Forest Institute.

Colantoni, A., Ferrara, C., Perini, L., & Salvati, L. (2015). Assessing trends in climate aridity and vulnerability to soil degradation in Italy. Ecological Indicators, 48, 599–604.

Corona, P., Ascoli, D., Barbati, A., Bovio, G., Colangelo, G., Elia, M., Garfì, V., Iovino, F., Lafortezza, R., Leone, V., Lovreglio, R., Marchetti, M., Marchi, E., Menguzzato, G., Nocentini, S., Picchio, R., Portoghesi, L., Puletti, N., Sanesi, G., & Chianucci, F. (2015). Integrated forest management to prevent wildfires under Mediterranean environments. Annals of Silvicultural Research, 39(1), 1–22.

Fares, S., Bajocco, S., Salvati, L., Camarretta, N., Dupuy, J.-L., Xanthopoulos, G., Guijarro, M., Madrigal, J., Hernando, C., & Corona, P. (2017). Characterizing potential wildland fire fuel in live vegetation in the Mediterranean region. Annals of Forest Science, 74, 1–14.

Fernandes, P. M., Rego, F. C., & Rigolot, E. (2011). The FIRE PARADOX project: Towards science-based fire management in Europe. Forest Ecology and Management, 261(12), 2177–2178.

Ferrara, A., Kelly, C., Wilson, G. A., Nolè, A., Mancino, G., Bajocco, S., & Salvati, L. (2016). Shaping the role of 'fast' and 'slow' drivers of change in forest–shrubland socio-ecological systems. Journal of Environmental Management, 169, 155–166.

Ferrara, C., Carlucci, M., Grigoriadis, E., Corona, P., & Salvati, L. (2017). A comprehensive insight into the geography of forest cover in Italy: Exploring the importance of socioeconomic local contexts. Forest Policy and Economics, 75, 12–22.

Ferretti, M., Marchetto, A., Arisci, S., Bussotti, F., Calderisi, M., Carnicelli, S., Cecchini, G., Fabbio, G., Bertini, G., Matteucci, G., de Cinti, B., Salvati, L., & Pompei, E. (2014). On the tracks of nitrogen deposition effects on temperate forests at their southern European range – An observational study from Italy. Global Change Biology, 20(11), 3423–3438.

Incerti, G., Feoli, E., Salvati, L., Brunetti, A., & Giovacchini, A. (2007). Analysis of bioclimatic time series and their neural network-based classification to characterise drought risk patterns in South Italy. International Journal of Biometeorology, 51, 253–263.

Kosmas, C., Detsis, V., Karamesouti, M., Kounalaki, K., Vassiliou, P., & Salvati, L. (2015). Exploring long-term impact of grazing management on land degradation in the socio-ecological system of Asteroussia Mountains, Greece. Land, 4(3), 541–559.

Mancino, G., Nolè, A., Salvati, L., & Ferrara, A. (2016). In-between forest expansion and cropland decline: A revised USLE model for soil erosion risk under land-use change in a Mediterranean region. Ecological Indicators, 71, 544–550.

Michetti, M., & Pinar, M. (2019). Forest fires across Italian regions and implications for climate change: A panel data analysis. Environmental and Resource Economics, 72(1), 207–246.

Moreira, F., Viedma, O., Arianoutsou, M., Curt, T., Koutsias, N., Rigolot, E., Barbati, A., Corona, P., Vaz, P., Xanthopoulos, G., Mouillot, F., & Bilgili, E. (2011). Landscape–wildfire interactions in southern Europe: Implications for landscape management. Journal of Environmental Management, 92(10), 2389–2402.

Nocentini, S., Buttoud, G., Ciancio, O., & Corona, P. (2017). Managing forests in a changing world: The need for a systemic approach. A review. Forest Systems, 26(1), eR01.

Pausas, J. G., Llovet, J., Rodrigo, A., & Vallejo, R. (2009). Are wildfires a disaster in the Mediterranean basin? A review. International Journal of Wildland Fire, 17(6), 713–723.

Raftoyannis, Y., Nocentini, S., Marchi, E., Calama, S. R., Garcia, G., Pilas, I., Peric, S., Amaral, P. J., Moreira-Marcelino, A. C., Costa-Ferreira, M., Kakouris, E., & Lindner, M. (2014). Perceptions of forest experts on climate change and fire management in the European Mediterranean forests. iForest, 7(1), 33–41.

Sallustio, L., Pettenella, D., Merlini, P., Romano, R., Salvati, L., Marchetti, M., & Corona, P. (2018). Assessing the economic marginality of agricultural lands in Italy to support land use planning. Land Use Policy, 76, 526–534.

Salvati, L., & Carlucci, M. (2011). The economic and environmental performances of rural districts in Italy: Are competitiveness and sustainability compatible targets? Ecological Economics, 70(12), 2446–2453.

Salvati, L., & Carlucci, M. (2014). A composite index of sustainable development at the local scale: Italy as a case study. Ecological Indicators, 43, 162–171.

Salvati, L., & Zitti, M. (2008). Regional convergence of environmental variables: Empirical evidences from land degradation. Ecological Economics, 68(1–2), 162–168.

Salvati, L., Perini, L., Sabbi, A., & Bajocco, S. (2012). Climate aridity and land use changes: A regional-scale analysis. Geographical Research, 50(2), 193–203.

Santopuoli, G., Ferranti, F., & Marchetti, M. (2015). Implementing criteria and indicators for sustainable forest management in a decentralized setting: Italy as a case study. Journal of Environmental Policy & Planning, 18(2), 177–196.

Seifollahi-Aghmiuni, S., Kalantari, Z., Egidi, G., Gaburova, L., & Salvati, L. (2022). Urbanisation-driven land degradation and socioeconomic challenges in peri-urban areas: Insights from Southern Europe. Ambio, 51(6), 1446–1458.

Tedim, F., Leone, V., & Xanthopoulos, G. (2016). A wildfire risk management concept based on a social-ecological approach in the European Union: Fire Smart Territory. International Journal of Disaster Risk Reduction, 18, 138–153.

Tomao, A., Quatrini, V., Corona, P., Ferrara, A., Lafortezza, R., & Salvati, L. (2017). Resilient landscapes in Mediterranean urban areas: Understanding factors influencing forest trends. Environmental Research, 156, 1–9.

Verkerk, P. J., Martinez de Arano, I., & Palahí, M. (2018). The bio-economy as an opportunity to tackle wildfires in Mediterranean forest ecosystems. Forest Policy and Economics, 86, 1–3.

Viedma, O., Moity, N., & Moreno, J. M. (2015). Changes in landscape fire-hazard during the second half of the 20th century: Agriculture abandonment and the changing role of driving factors. Agriculture, Ecosystems & Environment, 207, 126–140.

Viedma, O., Urbieta, I. R., & Moreno, J. M. (2018). Wildfires and the role of their drivers are changing over time in a large rural area of west-central Spain. Scientific Reports, 8(1), 17797.

Vilar, L., Gómez, I., Martínez-Vega, J., Echavarría, P., Riaño, D., & Martín, M. P. (2016). Multitemporal modelling of socio-economic wildfire drivers in Central Spain between the 1980s and the 2000s: Comparing generalized linear models to machine learning algorithms. PLOS ONE, 11(8), e0161344.

Vitolo, C., Di Giuseppe, F., Barnard, C., Coughlan, R., San-Miguel-Ayanz, J., Libertá, G., & Krzeminski, B. (2020). ERA5-based global meteorological wildfire danger maps. Scientific Data, 7(1), 216.

Vitolo, C., Di Giuseppe, F., Krzeminski, B., & San-Miguel-Ayanz, J. (2019). A 1980–2018 global fire danger re-analysis dataset for the Canadian Fire Weather Indices. Scientific Data, 6(1), 190032.

Wunder, S., Calkin, D. E., Charlton, V., Feder, S., Martínez de Arano, I., Moore, P., Rodríguez y Silva, F., Tacconi, L., & Vega-García, C. (2021). Resilient landscapes to prevent catastrophic forest fires: Socioeconomic insights towards a new paradigm. Forest Policy and Economics, 128, 102458.

Zuur, A. F., Ieno, E. N., & Elphick, C. S. (2010). A protocol for data exploration to avoid common statistical problems. Methods in Ecology and Evolution, 1(1), 3–14.

7

The Role of Local Landmarks in Environmental Education

Evangelos Papakitsos[1], Eleni Verouti[2], Marco Maialetti[3], Christina Papavasileiou[4], and Anastasios Mavrakis[5]

[1]Geo-Environmental Science and Environmental Quality Assurance Laboratory, University of West Attica, Greece
[2]Head of the Administration of Environment & Civil Protection at Municipality of Aspropyrgos, Greece
[3]Independent Researcher and Lawyer, Italy
[4]Second Athens Secondary Education Directorate, Greek Ministry of Education, Greece
[5]Secondary Education Directorate of West Attica, Greek Ministry of Education, Greece

Email: papakitsev@uniwa.gr; eleniverouti@yahoo.gr; verouti.e@aspropyrgos.gr; avv.marcomaialetti@gmail.com; xripapav@gmail.com; mavrakisan@yahoo.gr

Abstract

This work illustrates the pedagogical methodology employed, the students' response and engagement, as well as some key outcomes of the project, and the overall experience of the participating educators of an environmental education project entitled "Koumoundourou lake: a Habitat in our yard," integrated into a broader thematic context of "Human Rights in a Clean Environment and Culture" carried out in Greece. The initiative offered students an opportunity to explore a nearby yet often overlooked wetland ecosystem of both ecological and historical importance. Through active participation, research, and fieldwork, students developed critical environmental awareness, historical insight, and a personal connection to their local landscape.

DOI: 10.1201/9788743811107-7

7.1 Introduction

Environmental Education (EE) emerged as a core component of nature conservation and ecological policy in response to the growing concerns raised by natural scientists in the 1970s and 1980s regarding ecosystem degradation, pollution, and the depletion of natural resources (Palmos et al., 2021). One of the earliest conceptual models of EE, namely "Spaceship earth," emphasized the finite nature of earth's resources – excluding solar input – and promoted environmental stewardship as a survival imperative (Scoullos & Pavlidou, 2000). Over time, EE has evolved to encompass broader themes, including socioeconomic issues, historical and land-use dynamics, and international efforts toward sustainable development (Scoullos, 2004; Kousoulas, 2008; Palmos et al., 2021). Within this framework, the West Attica Directorate of Secondary Education in Greece supported the implementation of the project entitled "Koumoundourou Lake: A Habitat in Our Yard." The initiative was part of a wider EE program under the theme named "Human Rights in a Clean Environment and Culture."

Elefsina is located in the Saronic Gulf, southeast of the Greek mainland and approximately 24 kilometers from the port of Piraeus and downtown Athens, the capital city of Greece. The region presents a stark juxtaposition of natural beauty and industrial sprawl. Upon first encounter, one is confronted by massive industrial complexes – gas, steel, cement, and oil facilities – situated uncomfortably close to residential areas. Dust accumulates along roadways, particularly near the cement plants, yet protective measures appear absent: residents go about their lives unmasked, exposed to particulate matter (Scoullos, 2004). Amid this industrialized landscape, one must rely on imagination or historical knowledge to glimpse the lost grandeur of the Thriasio Plain (Figure 7.1).

Once home to fertile lands and pastoral traditions, the area is steeped in mythological and cultural heritage. Ancient Eleusis, now modern-day Elefsina, was among the most sacred cities of Classical Greece and the epicenter of the Eleusinian Mysteries – initiatory religious rites celebrating the goddesses Demeter and Persephone (Scoullos & Pavlidou, 2000).

7.2 The Environmental Burden of Industrialization

Despite its glorious past, Elefsina has come to symbolize environmental degradation. It has been identified as one of the most endangered regions in the Mediterranean and holds the record for one of the highest officially recorded

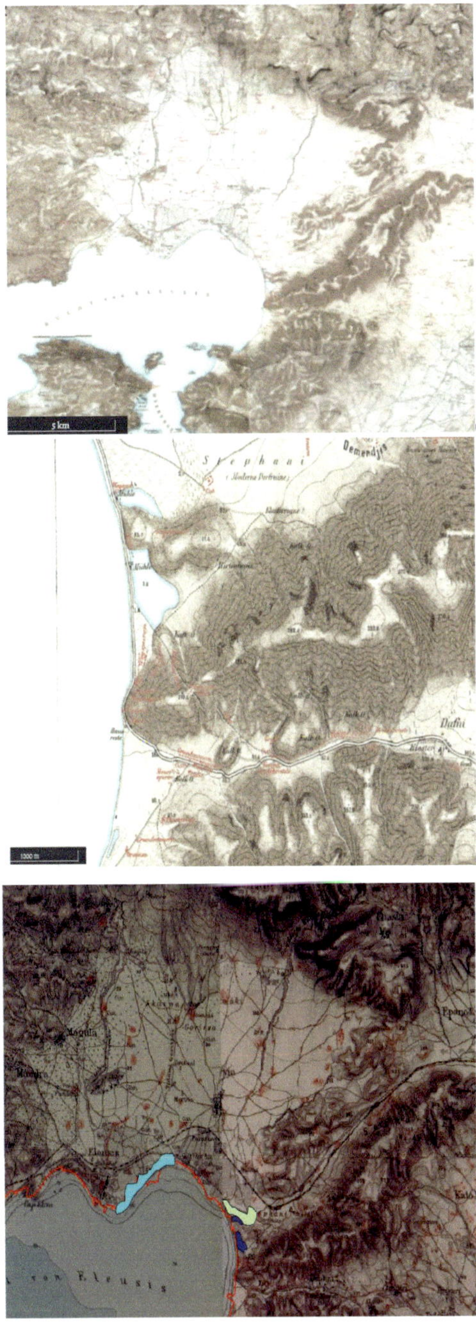

(a)

(b)

(c)

Figure 7.1 (a) "Koumoundourou" or "Rheitoi" lakes, as mapped in 1878 by J.A. Kaupert, through the website "Karten von Attika" via http://digi.ub.uni-heidelberg.de/diglit/curtius1900a and Curtius et al. (1881–1900); (b) Rheitoi lakes mapped in 1883 – the area of "Kalympaki" is highlighted; (c) analysis of the images gives us some information about the area, which cannot be noticed by the modern inhabitants, so in blue shading, the area of "Kalympaki" is highlighted, that is, the outflow of the aquifer, as well as the Lakes "Rheitoi" (Sakellaropoulos, 1884). Part of the Sacred Way passed east between "Rheitoi" lakes and Eastern hills (Archaeological Society Proceedings 1936, pp. 30, Figure 4); the Sacred Way passed between seashore and Koumoundourou lake as reported in the early 1930s (Proceedings of the Archaeological Society 1937, pp. 40, Figure 19); source: authors' elaboration of digital data sources available on the web.

temperatures in Europe (48.0 °C). This once-fertile valley, strategically located with maritime access to Corinth, the Peloponnese, and the Aegean, has been transformed into a dense industrial zone (Salvati & Carlucci, 2011). The industrial era began in the late 19th century, with the construction of the railway in 1884 and the opening of the Corinth Canal. These developments catalyzed commercial activities and population growth. Factories such as Charilaou Soap (1875), TITAN Cement (1902), VOTRYS, and KRONOS distilleries (early 20th century) followed rapidly.

The population increased further with the arrival of approximately 2000 refugees from Asia Minor in 1922, and subsequent waves of internal migration during the 1950s and 1960s (Papavasileiou & Mavrakis, 2013). Migrants from Epirus, the Dodecanese, Corfu, and Crete settled in Elefsina, transforming it from a coastal village into a bustling industrial hub. Today, the area hosts over 3000 industrial operations, including two oil refineries (Hellenic Petroleum and Petrola Hellas), steel industries, shipyards, chemical plants, ammunition manufacturers, scrap yards, and fuel depots (Papakitsos & Mavrakis, 2018; Palmos et al., 2021; Mavrakis et al., 2022).

The environmental toll of this industrial concentration is severe. Elefsina Bay suffers from degraded air and water quality due to emissions, fossil fuel consumption, vehicular traffic, and inadequate waste management (Salvati & Zitti, 2008). The coastal zone has experienced profound ecological damage, exacerbated by traffic congestion and noise pollution from heavy vehicles (Mavrakis et al., 2021; Mavrakis et al., 2022; Mavrakis et al., 2023). Industrial operations have contaminated air, soil, groundwater, and the marine environment. Elefsina serves as the main entry and processing point for crude oil in Greece (Imbrenda et al., 2021). Despite recent measures that have slightly improved water quality in the bay, pollutants – including chemical waste, hydrocarbons, and leachates – continue to affect the area (Christidis, 2007; Salvati et al., 2012, 2013). The situation typifies that of many urban-industrial port zones in Southern Europe (Seifollahi-Aghmiuni et al., 2022): unchecked urbanization, minimal enforcement of environmental regulations, and limited access to green and communal spaces (e.g., Colantoni et al., 2015; Salvati et al., 2015; Palmos et al., 2021).

7.3 Koumoundourou Lake: A Brief Description

Koumoundourou Lake – also known as the ancient Rheitoi – is the only natural lake in the region of Attica. Fed by freshwater springs from Mount

Figure 7.2 Kefalari, Koumoundourou (1943). Source: Georgios V. Tsingos' family.

Parnitha, the lake covers approximately 143,000 m^2 and features a mix of freshwater (60%) and seawater (40%) due to limited tidal exchange. A permanent freshwater discharge on its northern side maintains stable oxygen levels and helps preserve its aquatic life (Figure 7.2).

Despite its ecological potential, the lake has long suffered from environmental degradation. It lies in proximity to oil refineries (e.g., ELPE/ELDA), military fuel depots (e.g., AVEK Camp 871), and private fuel storage installations. Spills, leaks, and surface runoff from the nearby highway and industrial zone introduce high levels of pollutants into the lake (Kosmas et al., 2015). Measurements indicate chemical contamination levels that exceed safety thresholds by a factor of ten. Additionally, pesticides and fertilizers from surrounding agricultural activities contribute to eutrophication and toxicity (Christidis, 2007; Korres, 2008; Papavasileiou & Mavrakis, 2013; Markogianni et al., 2014; Papakitsos & Mavrakis, 2018). The Agricultural University of Athens intervened by raising the lake's water level by 20 cm to create a barrier against pollutant inflow (Poulovassilis et al., 1994). However, this was only a partial remedy; the lake remains heavily polluted and ecologically fragile (Paraschoudis, 2002).

7.3.1 Biodiversity and conservation status

The lake hosts over 50 species of birds, according to the Hellenic Ornithological Society. These include the globally threatened Ferruginous Duck (*Aythya nyroca*), as well as the Northern Shoveler, Pintail, Wigeon, and Pochard. Despite the severe pollution, the lake continues to offer habitat for avian species and remains a vital ecological niche in an otherwise industrialized region. Koumoundourou Lake was designated a historical site in 1974, recognizing its religious significance in antiquity. The ancient Sacred Way linking Athens to Eleusis passed directly beside the lake. Historical records and cartographic sources confirm that two lakes – collectively referred to as "Rheitoi" – existed in the area until the early 20th century. One of them, Kefalari (the northern lake), was drained during the construction of the oil refinery in the 1950s. Today, only one lake remains, partially encroached upon by the Athens–Corinth National Highway (Sakellaropoulos, 1884; Christidis, 2007; Markogianni et al., 2014).

7.3.2 A contested landscape

Although officially protected as an archaeological and suburban park, Koumoundourou Lake remained effectively isolated. Access was restricted under outdated military regulations dating from 1936 to 1953. Visitors had to report to the AVEK military camp and present identification to enter the area. A tangled mesh of barbed wire and warning signs surrounded much of the lake, reinforcing its invisibility from the public eye. Community groups and environmental organizations have called for the full reopening of the lake to the public, increased monitoring, and remediation efforts. The lake's resilience – despite industrial pollution and bureaucratic negligence – makes it an emblematic case of environmental survival and an ideal subject for experiential education. The students who participated in this project discovered not only a forgotten landscape but also a call to action.

The project's objective was to enable students to explore a proximate wetland of ecological, historical, and cultural significance. Koumoundourou Lake, a remnant of the ancient Rheitoi lakes, lies adjacent to Elefsina Bay, amidst one of Greece's most industrialized and environmentally challenged regions. Despite its proximity, the lake remains unfamiliar and inaccessible to most locals and residents. The project aimed to address this disconnect by transforming the lake from a hidden environmental landmark into an educational catalyst. Through voluntary participation, students actively engaged in the investigation of the lake's past and present.

Their work fostered a deeper understanding of the ecosystem, enhanced their environmental consciousness, and reinforced their right to a healthy environment and cultural continuity. This paper outlines the methodological approach, pedagogical rationale, student-led activities, and the transformative outcomes observed during the project. In highlighting Koumoundourou Lake as both a threatened habitat and a symbol of resilience, we argue for the critical role of local landmarks in contextual environmental education.

7.4 Teaching Approach

The educational program adopted a participatory, inquiry-based approach that encouraged students to collaborate and engage directly with both primary and secondary sources. The core group consisted of 25 students, divided into sub-groups of five. Each group was assigned a specific sub-topic related to Koumoundourou Lake and the overarching theme of environmental rights. The project implementation included: (i) classroom lectures and group discussions, (ii) web-based and bibliographic research, (iii) field exploration and photo documentation, (iv) preparation of final reports, PowerPoint presentations, and a thematic poster, and (v) a structured field visit to Koumoundourou Lake (Papavasileiou and Mavrakis, 2013). Sub-groups examined thematic areas such as the Universal Declaration of Human Rights, emphasizing the right to a clean and safe environment; the ecological, geographical, and historical significance of Koumoundourou Lake; the impact of pollution and industrialization on the local ecosystem, as well as cultural and archaeological references to the Sacred Way and the ancient Rheitoi lakes.

Students used web-based research strategies to collect data, including historical maps, archival photos, and scientific articles (Figure 7.3). Emphasis was placed on experiential learning, with students encouraged to investigate and reflect on their findings through discussion and analysis (Demetriou et al, 2008; Markogianni et al., 2014; Papakitsos and Mavrakis, 2018). Aligned with the Greek Pedagogical Institute's guidelines on Environmental Education (Kousoulas, 2008), the project aimed at promoting sustainable development awareness, enhancing collaborative learning and teamwork, and fostering critical and analytical thinking.

Students were expected to:

- Identify and appreciate the natural, historical, and cultural value of Koumoundourou Lake.
- Understand the ecological significance of wetlands and the mechanisms of environmental degradation.

Figure 7.3 A partial documentation of extreme events in the area. (a), (b) (c) 08-07-2018 in front of the Lake Koumoundourou, the flooded National Road Athens Elefsis. (d) Daily accumulated rainfall at meteorological stations, maximum precipitation of 48.4 mm during July 8, 2018; source: authors' archive and the National Observatory of Athens.

- Strengthen their connection with local heritage and develop environmental responsibility.
- Cultivate the motivation to advocate for ecological preservation.
- Recognize the broader implications of pollution and urbanization (Demetriou et al, 2008).

7.5 Project Implementation

Students conducted in-depth research using multiple sources, including (1) Kaupert's 1878 map, which depicted the original twin lakes, and (2) photographs from the early 20th century, including 1936 and 1937 documentation of the Sacred Way and lakeshore, as well as (3) Reports by Scoullos & Pavlidou

(2000) and Demetriou et al. (2008), detailing pollution levels and hydrogeological data. These resources enabled students to understand land-use changes and the lake's transformation due to industrial development and the construction of the Athens–Corinth National Highway (*sensu* Ferrara et al., 2016).

The field trip to Koumoundourou Lake was both revealing and emotional. Despite the lake's proximity to a noisy highway and industrial facilities, students observed vibrant biodiversity, including birds and wild plants. They expressed surprise at discovering such a rich ecosystem hidden within their familiar surroundings. Many students enthusiastically documented the site with their cameras. For most, this was a first encounter with a natural wetland and its resident species. The visit transformed the lake from an abstract environmental issue into a living, breathing place (Figure 7.4).

Based on intense desk and field work, students concluded that Koumoundourou Lake is the only natural lake in Attica, measuring 143,000 m^2, with a shoreline of 1680 meters and an average depth of 1.5 meters. Water input includes springs, seawater intrusion, and rainfall, with a weir system regulating outflow toward the bay (Tomao et al., 2017). The lake's deterioration results from industrial effluents, oil leaks, urban runoff, pesticides, and lack of modern infrastructure (e.g., Jakubínský et al., 2021). Despite degradation, the lake retains biodiversity and ecological value, documented through bird sightings and plant identification (e.g., Incerti et al., 2007). Historical mapping and photography illustrated the continuity of the site's cultural and environmental role from antiquity to today (Salvati & Zitti, 2005). Building on such assumptions, students published their findings on their school's website and the West Attica Directorate of Secondary Education's platform (Palmos et al., 2021).

7.6 Concluding Remarks

The Environmental Education project at the 1st Technical High School of Aspropyrgos successfully achieved its goals. It exposed students to a local ecosystem of immense ecological, cultural, and historical value. Through research and field experience, students (i) gained critical insights into environmental challenges, (ii) discovered the importance of Koumoundourou Lake as a habitat and heritage site, (iii) understood the effects of industrial development and land-use change, and (iv) developed a personal and collective sense of environmental responsibility (e.g., Wilson et al., 2017). Koumoundourou Lake remains a resilient but vulnerable wetland, symbolizing the complex relationship between nature, culture, and development. Its preservation is not only a scientific imperative but also a civic and moral one.

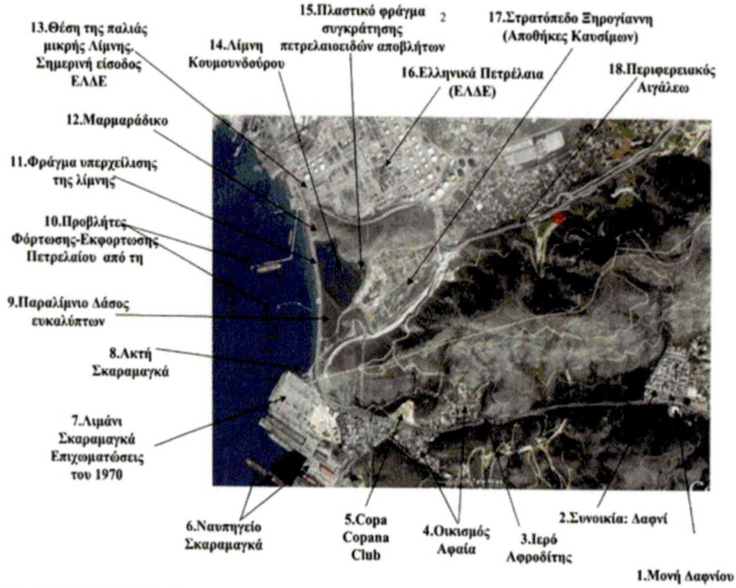

(a)

(b)

Figure 7.4 (a) Picture of the worksheet given to students for field research (Koumoundourou Lake). Students were forced to recognize places, installations, and environmental characteristics. 1) Daphni Monastery, 2) Daphni neighborhood, 3) Temple of Aphrodite, 4) Aphaia settlement, 5) Copa Copana Club, 6) Skaramanga Shipyard, 7) Skaramanga Shipyard. Landfill in 1970, 8) Skaramanga Coast, 9) Paralimnio Eucalyptus Forest, 10) Petroleum Loading and Unloading Piers, 11) Lake Overflow Dam, 12) Marmaradiko, 13) Location of the Old Small Lake. Current entrance to ELDA (ELPE), 14) Koumoundourou Lake, 15) Plastic dam for retaining oil waste, 16) Hellenic Petroleum, 17) Xirogiannis Camp – Fuel Storage Facilities, 18) Aegaleo regional road. (b) Screenshot dated 2025-08-07 at 23:53:20, Regional Unit of East Attica, Greece, extracted on August 04, 2025 (Satellite Imagery). Source: authors' elaboration on digital archives.

References

Christidis, A. (2007). Annual reports of water and soil chemical analyses of Koumoundourou Lake (1984–2006) (in Greek). Bureau of Pollution Control and Environmental Quality of the Development Association of Thriassion Plain, Aspropyrgos.

Colantoni, A., Ferrara, C., Perini, L., & Salvati, L. (2015). Assessing trends in climate aridity and vulnerability to soil degradation in Italy. Ecological Indicators, 48, 599-604.

Curtius, E., Kaupert, J. A., & Milchhoeffer, A. (1881–1900). Karten von Attika. Berlin. Retrieved from [http://digi.ub.uni-heidelberg.de/diglit/curtius1900a] accessed June 2025.

Demetriou, E., Karaouzas, I., Sarantakos, K., Zacharias, I., Bogdanos, K., & Diapoulis, A. (2008). Groundwater risk assessment at a heavily industrialised catchment and the associated impacts on a peri-urban wetland. Journal of Environmental Management, 88, 526-538.

Ferrara, A., Kelly, C., Wilson, G. A., Nolè, A., Mancino, G., Bajocco, S., & Salvati, L. (2016). Shaping the role of 'fast' and 'slow' drivers of change in forest-shrubland socio-ecological systems. Journal of Environmental Management, 169, 155-166.

Ferrara, A., Kosmas, C., Salvati, L., Padula, A., Mancino, G., & Nolè, A. (2020). Updating the MEDALUS ESA framework for worldwide land degradation and desertification assessment. Land Degradation & Development, 31(12), 1593-1607.

Francaviglia, R., Di Bene, C., Farina, R., Salvati, L., & Vicente-Vicente, J. L. (2019). Assessing "4 per 1000" soil organic carbon storage rates under Mediterranean climate: A comprehensive data analysis. Mitigation and Adaptation Strategies for Global Change, 24, 795-818.

Imbrenda, V., Quaranta, G., Salvia, R., Egidi, G., Salvati, L., Prokopová, M., Coluzzi, R., & Lanfredi, M. (2021). Land degradation and metropolitan expansion in a peri-urban environment. Geomatics, Natural Hazards and Risk, 12(1), 1797-1818.

Incerti, G., Feoli, E., Salvati, L., Brunetti, A., & Giovacchini, A. (2007). Analysis of bioclimatic time series and their neural network-based classification to characterise drought risk patterns in South Italy. International Journal of Biometeorology, 51, 253-263.

Jakubínský, J., Prokopová, M., Raška, P., Salvati, L., Bezak, N., Cudlín, O., Cudlín, P., Purkyt, J., Vezza, P., Camporeale, C., Daněk, J., Pástor, M., & Lepeška, T. (2021). Managing floodplains using nature-based solutions to support multiple ecosystem functions and services. Wiley Interdisciplinary Reviews: Water, 8(5), e1545.

Korres, M. (2008). Maps of Attica by Ernest Kourtiou & Ioannis A. Kaoupert, Melissa, Athens.

Kosmas, C., Detsis, V., Karamesouti, M., Kounalaki, K., Vassiliou, P., & Salvati, L. (2015). Exploring long-term impact of grazing management on land degradation in the socio-ecological system of Asteroussia Mountains, Greece. Land, 4(3), 541–559.

Kousoulas, G. K. (2008). Approaching Environmental Education. Technical Report, LIFE05 ENV/GR/000245. National Center for Social Research, Institute of Urban and Rural Sociology. Available at [www.ekke.gr/estia] accessed June 2025.

Markogianni, V., Dimitriou, E., & Karaouzas, I. (2014). Water quality monitoring and assessment of an urban Mediterranean lake facilitated by remote sensing applications. Environmental Monitoring and Assessment, 186, 5009–5026.

Mavrakis, A., Tsigkou, A., Papavasileiou, C., Sigalos, G., Dasaclis, S., Vamvakeros, X., & Papakitsos, C. E. (2022). An application of Systems Science in the context of Greek environmental education for planning interventions against volcanic hazards. International Journal of Research in Educational Sciences, 5(3), 41–76.

Mavrakis, A., Lykoudis, S., & Salvati, L. (2023). Predicting the occurrence of natural and technological disasters in Greece through Verhulst, multinomial and exponential models. Safety Science, 166, 106246.

Mavrakis, A., Papavasileiou, C., Alexakis, D., Papakitsos, C. E., & Salvati, L. (2021). Meteorological patterns and the evolution of West Nile virus in an environmentally stressed Mediterranean area. Environmental Monitoring and Assessment, 193(4), 227.

Palmos, D., Papavasileiou, C., Papakitsos, C. E., Vamvakeros, X., & Mavrakis, A. (2021). Enhancing the environmental programmes of secondary education by using web-tools concerning precaution measures in civil protection: The case of Western Attica (Greece). Safety Science, 135, 105117.

Papakitsos, E. C., & Mavrakis, A. (2018). A systemic model proposed for the management of local environmental education, awareness and protection: A case study. Humanities and Social Science Research, 1(2), 1–8.

Papavasileiou, H., & Mavrakis, A. (2013). Environmental education: Issue water: Different approaches in secondary general and technical lyceum in a social and environmental stressed area in Greece. Procedia Technology, 8, 171-174.

Paraschoudis, V. (2002). Hydrogeological study of Western Attica (Megara–Thriasio Plains). Hellenic Ministry of Agriculture, Department of Geology-Hydrology, Athens.

Poulovassilis, A., Mimides, T., Faltseta, S., Gianopoulos, P., & Sgoumpopoulou, A. (1994). Hydrogeology of Lake Koumoundourou. Pollution sources and its protection. Agronomic University of Athens.

Proceedings of the Archaeological Society. (1936). p. 30, fig. 4.

Proceedings of the Archaeological Society. (1937). p. 40, fig. 19.

Sakellaropoulos, D. K. (1884). The maps of Attica (Karten von Attika). Apollon Magazine, B(16), 241–244.

Salvati, L., & Carlucci, M. (2011). The economic and environmental performances of rural districts in Italy: Are competitiveness and sustainability compatible targets? Ecological Economics, 70(12), 2446–2453.

Salvati, L., & Zitti, M. (2005). Land degradation in the Mediterranean basin: Linking bio-physical and economic factors into an ecological perspective. Biota, 6, 67–77.

Salvati, L., & Zitti, M. (2008). Regional convergence of environmental variables: Empirical evidences from land degradation. Ecological Economics, 68(1-2), 162–168.

Salvati, L., Mavrakis, A., Colantoni, A., Mancino, G., & Ferrara, A. (2015). Complex adaptive systems, soil degradation and land sensitivity to desertification: A multivariate assessment of Italian agro-forest landscape. Science of the Total Environment, 521, 235–245.

Salvati, L., Perini, L., Sabbi, A., & Bajocco, S. (2012). Climate aridity and land use changes: A regional-scale analysis. Geographical Research, 50(2), 193–203.

Salvati, L., Tombolini, I., Perini, L., & Ferrara, A. (2013). Landscape changes and environmental quality: The evolution of land vulnerability and potential resilience to degradation in Italy. Regional Environmental Change, 13, 1223–1233.

Scoullos, M. (2004). Concepts and international initiatives in environmental education. In: Proceedings of the 2nd National Conference on Sustainable Development and Environmental Education, Halkidiki, Greece.

Scoullos, M. J., & Pavlidou, A. S. (2000). Metal speciation studies in a brackish/marine interface system. Global NEST Journal, 2(3), 255–264.

Seifollahi-Aghmiuni, S., Kalantari, Z., Egidi, G., Gaburova, L., & Salvati, L. (2022). Urbanisation-driven land degradation and socioeconomic challenges in peri-urban areas: Insights from Southern Europe. Ambio, 51(6), 1446–1458.

Tomao, A., Quatrini, V., Corona, P., Ferrara, A., Lafortezza, R., & Salvati, L. (2017). Resilient landscapes in Mediterranean urban areas: Understanding factors influencing forest trends. Environmental Research, 156, 1–9.

Wilson, G. A., Kelly, C. L., Briassoulis, H., Ferrara, A., Quaranta, G., Salvia, R., Detsis, V., Curfs, M., Cerda, A., El-Aich, A., Liu, H., Kosmas, C., Alados, C. L., Imeson, A., Landgrebe-Trinkunaite, R., Salvati, L., Naumann, S., Danwen, H., Iosifides, T., Kizos, T., Mancino, G., Nolè, A., Jiang, M., & Zhang, P. (2017). Social memory and the resilience of communities affected by land degradation. Land Degradation & Development, 28(2), 383–400.

8

Unraveling the Inherent Complexity in Spatio-temporal Patterns of Urbanization: Theoretical and Empirical Contributions from Global to Local Observation Scales

Marco Maialetti[1], Clio Ciaschini[2], Donato Scarpitta[3], Bogdana Nosova[4], Gianluigi Salvucci[5], and Luca Salvati[6]

[1]Independent Researcher and Lawyer, Italy
[2]Department of Economics and Social Sciences, Polytechnic University of Marche, Italy
[3]Independent Researcher and Architect, Italy
[4]Department of Social Communications, Institute of Journalism, Taras Shevchenko National University of Kyiv, Ukraine
[5]Italian National Institute of Statistics (ISTAT), Italy
[6]Department of Methods and Models for Economics, Territory and Finance (MEMOTEF), Faculty of Economics, Sapienza University of Rome, Italy

Email: avv.marcomaialetti@gmail.com; c.ciaschini@univpm.it; scarpitta.donato@libero.it; bogdana.nosova@gmail.com; salvucci@istat.it; luca.salvati@uniroma1.it

Abstract

When pursued over extended temporal horizons, the systematic examination of demographic dynamics offers a privileged perspective to envision complex socioeconomic processes underlying the genesis and persistence of territorial disparities and inequalities at the landscape scale. Nevertheless, notwithstanding the centrality of such an approach, empirical inquiries into the long-term role of population density in structuring human settlement systems along urban–rural gradients – whether within regions, across nations, or at continental scales – remain comparatively limited, even in world areas

DOI: 10.1201/9788743811107-8

marked by an exceptionally deep settlement history, such as Europe. In recent years, selected investigations have endeavored to disentangle these dynamics through a set of case studies ranging from the global to the European and national levels, making use of advanced data mining procedures and spatial econometric models – both cross-sectional and longitudinal – in their global, quantile, and local formulations. The empirical evidence of recent studies illuminates the extent to which density- and path-dependent mechanisms correspond to specific phases within broader economic and urban cycles. In addition, the investigation foregrounds the intricate interplay between economic forces – such as agglomeration dynamics and economies of scale – and non-economic dimensions, including institutional arrangements, cultural specificities, and territorial structures, all situated within a historically informed regional perspective. Through the interpretation of spatio-temporal economic configurations and socio-demographic indicators, recent studies ultimately aspire to inform the design of regional development strategies and spatial planning policies capable of responding with flexibility to heterogeneous and rapidly evolving local contexts.

8.1 Introduction

Notwithstanding their acknowledged significance in developmental research, systematic and comparative analyses of divergent trajectories of local development remain relatively limited across diverse world regions (Scheuer et al., 2016). Examining demographic trajectories at the local scale, within geographical contexts where individual countries reflect distinctive historical, socioeconomic, political, and cultural legacies, offers privileged insights into the latent mechanisms underpinning regional growth (Kabisch & Haase, 2011). In advanced economies, such trajectories have frequently mirrored periods of profound economic restructuring, giving rise to urbanization–suburbanization–reurbanization sequences, coupled with broader demographic transitions characterized by declining fertility, the postponement of childbearing, rising life expectancy, and sustained immigration flows (Lopez-Gay, 2014).

Europe, in particular, has experienced especially intricate urban cycles in the post-war period, with dispersed urbanization increasingly stimulating suburban residential mobility. Analyses conducted at disaggregated spatial scales reveal underlying tendencies toward short-distance mobility and a growing preference for larger dwellings in peri-urban contexts. A plurality of structural factors shapes the spatial distribution of resident populations in advanced economies, encompassing globalization, the transformation of economic systems, and international migration (Cheshire, 1995).

Yet density-dependent mechanisms of population growth continue to exert a decisive influence on demographic dynamics at the regional scale (Champion, 2001). Earlier contributions have substantiated the hypothesis of density-dependent regulation of human population size (Bell, 2015), while simultaneously emphasizing its divergence from the processes that govern non-human populations (Duncan et al., 2001). Unlike in animal ecology, density-dependent regulation accounts only partially for the observed variability in human population growth rates (Hamilton et al., 2009). They are intrinsically mediated by individual preferences and choices, cultural, ethnic, and religious specificities, broader socioeconomic conditions, and exogenous stochastic processes that are not always identifiable or readily amenable to formal modeling (Millward, 2008).

Nevertheless, notable parallels with ecological literature remain apparent: (i) the intrinsic stochasticity of density-dependent mechanisms, manifesting in the predominance of positive, neutral, or negative effects of density on short-term growth rates; (ii) their temporal volatility; and (iii) their spatial heterogeneity, which collectively generate the intricate mosaic of expansion and decline characteristic of human populations (e.g., Serra et al., 2014; Chelli et al., 2016; Quaranta et al., 2016). These dynamics are most clearly observable under conditions of demographic equilibrium, over extended temporal horizons and within spatial frameworks sufficiently broad to capture short-term and medium-range population movements – allowing for the differentiation of density-dependent from density-independent regulatory processes (Cudlin et al., 2017; Osterhage, 2018; Benassi & Naccarato, 2019).

Within this framework, the Northern Mediterranean basin emerges as a paradigmatic case of urban expansion situated at the intersection of advanced and emerging socioeconomic dynamics (Figure 8.1).

Resident population in the region expanded from approximately 90 million in the early 1950s to 260 million by the late 1990s and is projected to reach 415 million by 2030. Urbanization rates, historically ranging from 60% in Greece to 80% in Spain, are expected to increase markedly by that date. The settlement history of Southern European countries thus exemplifies heterogeneous local-scale demographic dynamics, distinct both in population size and structural features when compared with other European and extra-European realities (Chtouris et al., 2006). Densest population clusters are found in central cities and their metropolitan areas, encompassing capitals and major urban centers (Zambon et al., 2018).

Based on these premises, mechanisms of urban–rural expansion and decline represent an especially compelling research frontier. It calls for multidisciplinary approaches and poses significant challenges to official statistics, policy design, and the pursuit of sustainable regional development, social

Figure 8.1 Suburban settlements in-between compactness and dispersion, typical of large metropolitan regions in the Mediterranean; source: authors' photographic archive.

planning, and territorial governance. The present work sheds further light on this scientific topic, providing a brief commentary on complex and latent mechanisms shaping urbanization patterns and processes and thus influencing both morphology and functions of landscapes at the interface between urban and rural settlements.

8.2 Monitoring Latent Urbanization Processes: An Operational Framework

A comparative analysis of the historical evolution of population distribution and density can be run exploiting the potential of official statistics at different spatial scales (global, continental, national) and encompassing different temporal scales. The operational framework can be structured into three main activities. First, a literature review and collection of historical evidence based on quantitative data and qualitative information can be carried out, justifying and corroborating the results of statistical analysis over time and space, and focusing on the intrinsic peculiarity of different case studies in a purely geo-historical perspective. The second step should include a data collection phase, implying homogenization and preparation for data mining and econometrics with separate approaches and technical solutions for each of the three case studies. Multiple databases and other information sources can be adopted, considering a kaleidoscope of different examples and case studies, as follows:

(i) A global dataset provided by the United Nations and quantifying population dynamics between 1950 and 2020 in nearly 1800 metropolitan areas (>300,000 inhabitants in 2014) of more than 150 world countries, with annual temporal resolution;

(ii) A continental dataset provided by Eurostat and quantifying population dynamics in each municipality of 27 European countries (more than 500,000 observations per year) between 1960 and 2010, harmonized from general demographic censuses with decadal temporal resolution;

(iii) Selected national datasets covering three representative Mediterranean countries in Europe with important spatial disparities (e.g., Italy, Spain, and Greece), exploring municipal spatial domains and encompassing longer time windows.

Estimates of resident population can be made available at the lowest available spatial domain (cities in the case of (i), municipalities in the case of (ii) and (iii)), approximately every 10 years over the longest time interval available. Representing the dependent variable in most econometric models, population growth rates can be calculated at each spatial domain over coherent time intervals of similar length, depending on data availability. Population density (inhabitants/km^2) can be calculated at the same time points and spatial scale. In all case studies, demographic data will refer to homogenized geographical boundaries within the study period. Other (specific) predictors can be made available on a case-by-case basis, depending on data availability.

Examples of possible predictors (or contextual variables) may include vital statistics, and thus, natural balance, migratory balance, infant mortality rate, marriage rate, urbanization rate, and elevation, providing a context-based analysis of demographic trends, accessibility, and urbanization processes. Statistical tests should be extensively used in order to keep predictors' redundancy to a minimum and to manage multi-collinearity, using spectral decomposition in (independent) principal components or instrumental variables, when necessary. Tools and facilities of geographical information systems (GIS) may also support data collection, variable homogenization, and indicators' calculation (Franch-Pardo et al., 2020). In case studies (ii) and (iii), considering municipalities (or local districts) as the elementary analysis unit allows exploiting a relatively vast (and often underexplored) information. Based on these data sources, mechanisms of population growth and decline can be identified in each case mentioned above from (i) to (iii), using a wide set of exploratory multivariate statistical techniques (data mining) and spatially explicit econometric approaches (both cross-section and panel).

8.2.1 Quantitative approaches

To study the (short-term) evolution of population dynamics over distinctive stages of the metropolitan cycle, cross-section regressions can be adopted, whose results should be contrasted with the empirical findings of panel regressions, specifically run when evaluating the overall, long-term urbanization dynamics (e.g., Salvati et al., 2016). While presenting a variable goodness-of-fit, consistent regression outputs (i.e., the same significant predictors with comparable intensity and sign) identify a statistically stable (and conceptually relevant) relationship between population growth rates and the selected predictors, under a specific background context (i.e., a given phase of the metropolitan cycle). Econometric tests may be extensively used to define optimal specifications under a comparative framework. Based on earlier evidence from an extensive literature review, a (global) spatial Durbin model (SDM) can be adopted as a baseline, considering three spatial matrices (Queen contiguity, *knn* (defined order), inverse distance weighting), modeling short-haul, medium-haul, and long-haul spatial interactions among predictors and the dependent variable, respectively.

Additional specifications could be run under specific results of econometric tests, indicating non-optimality of the models initially run. In this perspective, geographically weighted regressions (GWRs) estimate cross-section (separately for each stage of the metropolitan cycle) and under a panel framework (to study the aggregate effect on population growth). The results of exploratory approaches may integrate econometrics – and are possibly interpreted (and finely integrated) with an extensive literature review and geo-historical investigation designed for each case study (e.g., Lamonica et al., 2020). This aims at delineating the role of economic factors (e.g., agglomeration and scale) and non-economic factors (demographic, institutional, and cultural forces).

8.3 Urbanization and Landscape Evolution: Discussing Positive and Normative Implications

The rigorous interpretation of economic data and socio-demographic indicators through statistical and econometric modeling (Colantoni et al., 2021), combined with a historical-geographical perspective grounded in an extensive literature review and supplemented by additional qualitative data sources, represents a novel contribution to applied economics, regional science, and spatial demography. The empirical insights generated by this approach are expected to elucidate the ways in which density-dependent and path-dependent mechanisms are closely associated with specific stages of metropolitan cycles. Density-growth relationships can be interpreted as reflective of the sequential waves characteristic of metropolitan life cycles. Recent literature increasingly highlights the persistent role of density-dependence and path-dependence in shaping demographic dynamics at local, regional, and national scales (Salvati & Zitti, 2009). This constitutes a novel, multidisciplinary contribution spanning a wide range of fields, from applied economics and regional studies to spatial demography, human geography, planning, and sustainability science (Salvati & Carlucci, 2011).

Moreover, the analysis of density-dependent and path-dependent patterns of population growth and decline over extended, historically relevant time intervals permits a more nuanced understanding of the socioeconomic processes underlying territorial disparities, providing both scientific insight and normative guidance for policy interventions. An additional innovation of this study lies in its operational approach, which integrates quali-quantitative methodologies with a combination of parametric and non-parametric statistical techniques for comprehensive data analysis. Assessing density-dependence, path-dependence, agglomeration and scale effects, as well as spatial interactions across the full spectrum of the metropolitan cycle – from urbanization to re-urbanization – is a task of particular urgency and relevance (Egidi et al., 2020).

Testing density-dependent population growth over time contributes to delineating and comparing the mechanisms that underpin metropolitan hierarchies' representative of both advanced and emerging economies (Salvati et al., 2013). Simultaneously, density-dependence is considered foundational for understanding local-scale settlement evolution in rural areas, a phenomenon common across diverse national contexts. Looking ahead, a particularly compelling research agenda involves documenting how sequential waves of urban and rural concentration and deconcentration interact with population growth dynamics (Salvati, 2013).

8.3.1 Urbanization, landscape, sustainability: Three "inclusive" hegemonic concepts

From a normative perspective, this research integrates quantitative and qualitative methodologies to empirically investigate path-dependence across regions, countries, and continents. Such an inquiry constitutes a particularly relevant endeavor for sustainability science, especially in light of the objectives of the 2030 Agenda for Sustainable Development, with specific reference to Sustainable Development Goals 8, 10, 11, and 15 (Salvati & Carlucci, 2014). A refined understanding of the complex spatio-temporal patterns of population dynamics – and of the underlying socioeconomic mechanisms driving demographic growth and decline – offers critical insights for rethinking spatial planning (Munafò et al., 2010). From the operational perspective of official statistics, a more nuanced analysis of socioeconomic contexts emerging from diverse territorial patterns and processes enhances the reliability and precision of population forecasts at medium- and fine-grained spatial resolutions – a particularly innovative contribution to regional science, applied economics, and spatial demography (Seifollahi-Aghmiuni et al., 2022). In this regard, the operational approach proposed in this study underscores the pivotal role of geo-referenced official databases that provide local-scale, up-to-date information spanning historically significant time intervals (Tomao et al., 2017). Furthermore, it encourages the adoption of spatially explicit econometric frameworks in the analysis of population dynamics and urban growth (e.g., Sallustio et al., 2018).

8.4 Concluding Remarks

Within this framework, the present study debates the historical evolution of population distribution and density using authoritative statistical sources. The proposed processes of data collection, harmonization, and the construction of relevant indicators ensure consistency and reliability in the final analysis. A quantitative approach, combining global techniques – both parametric and non-parametric quantile methods – and locally oriented models such as geographically weighted regression, is proposed to achieve basic research objectives that answer common research assumptions. Indicators derived from such models, including measures of fit and regression coefficients, provide a comprehensive assessment of the influence of regulatory mechanisms on population dynamics across both time and space. By integrating these empirical results with basic socioeconomic indicators and insights drawn from a wide-ranging review of historical literature, this contribution aims to uncover the underlying causes of spatial disparities, highlighting long-term and more recent socioeconomic trends and their effects on settlement patterns and metropolitan development under a "complex landscape systems" perspective.

References

Alados, C.L., Errea, P., Gartzia, M., Saiz, H., & Escós, J. (2014). Positive and negative feedbacks and free-scale pattern distribution in rural-population dynamics. PloS One, 9, 12.

Bell, M. (2015). Demography, time and space. Journal of Population Research, 32(3–4), 173–186.

Benassi, F., & Naccarato, A. (2019). Modelling the spatial variation of human population density using Taylor's power law, Italy, 1971–2011. Regional Studies, 53(2), 206–216.

Champion, A.G. (2001). A changing demographic regime and evolving polycentric urban regions: Consequences for the size, composition and distribution of city populations. Urban Studies, 38(4), 657–677.

Chelli, F.M., Ciommi, M., Emili, A., Gigliarano, C., & Taralli, S. (2016). Assessing the equitable and sustainable well-being of the Italian provinces. International Journal of Uncertainty, Fuzziness and Knowledge-Based Systems, 24(Suppl. 1), 39–62.

Cheshire, P. (1995). A new phase of urban development in Western Europe? The evidence for the 1980s. Urban Studies, 32(7), 1045–1063.

Chtouris, S., Zissi, A., Papanis, E., & Rontos, K. (2006). The state of youth in contemporary Greece. Young, 14(4), 309–322.

Cohen, J.E. (2003). Human population: the next half century. Science, 302, 1172–1175.

Colantoni, A., Villarini, M., Monarca, D., Carlini, M., Mosconi, E.M., Bocci, E., & Hamedani, S.R. (2021). Economic analysis and risk assessment of biomass gasification CHP systems of different sizes through Monte Carlo simulation. Energy Reports, 7, 1954–1961.

Coluzzi, R., Bianchini, L., Egidi, G., Cudlin, P., Imbrenda, V., Salvati, L., & Lanfredi, M. (2022). Density matters? Settlement expansion and land degradation in Peri-urban and rural districts of Italy. Environmental Impact Assessment Review, 92, 106703.

Cudlin, P., Klopčič, M., Tognetti, R., Máli, F., Alados, C.L., Bebi, P., & Wielgolaski, F.E. (2017). Drivers of treeline shift in different European mountains. Climate Research, 73(1-2), 135–150.

Duncan, S.R., Duncan, C.J., & Scott, S. (2001). Human population dynamics. Annals of Human Biology, 28(6), 599–615.

Egidi, G., Salvati, L., & Vinci, S. (2020). The long way to tipperary: City size and worldwide urban population trends, 1950–2030. Sustainable Cities and Society, 60, 102148.

Franch-Pardo, I., Desjardins, M. R., Barea-Navarro, I., & Cerdà, A. (2021). A review of GIS methodologies to analyze the dynamics of COVID-19 in the second half of 2020. Transactions in GIS, 25(5), 2191–2239.

Hamilton, M.J., Burger, O., DeLong, J.P., Walker, R.S., Moses, M.E., & Brown, J.H. (2009). Population stability, cooperation, and the invasibility of the human species. Proceedings of the National Academy of Sciences USA, 106, 12255–12260.

Kabisch, N., & Haase, D. (2011). Diversifying European agglomerations: evidence of urban population trends for the 21st century. Population, Space and Place, 17(3), 236–253.

Lamonica, G. R., Recchioni, M. C., Chelli, F. M., & Salvati, L. (2020). The efficiency of the cross-entropy method when estimating the technical coefficients of input–output tables. Spatial Economic Analysis, 15(1), 62–91.

López-Gay, A. (2014). Population growth and re-urbanization in Spanish inner cities: The role of internal migration and residential mobility. Quetelet Journal, 1(2), 67–92.

Millward, H. (2008). Evolution of population densities: five Canadian cities, 1971-2001. Urban Geography, 29, 616–638.

Munafo, M., Norero, C., Sabbi, A., & Salvati, L. (2010). Soil sealing in the growing city: a survey in Rome, Italy. Scottish Geographical Journal, 126(3), 153–161.

Osterhage, F. (2018). The End of Reurbanisation? Phases of Concentration and Deconcentration in Migratory Movements in North Rhine-Westphalia. Comparative Population Studies, 43, 131–156.

Quaranta, G., Citro, E., & Salvia, R. (2016). Economic and social sustainable synergies to promote innovations in rural tourism and local development. Sustainability, 8(7), 668.

Sallustio, L., Pettenella, D., Merlini, P., Romano, R., Salvati, L., Marchetti, M., & Corona, P. (2018). Assessing the economic marginality of agricultural lands in Italy to support land use planning. Land Use Policy, 76, 526–534.

Salvati, L. (2013). Monitoring high-quality soil consumption driven by urban pressure in a growing city (Rome, Italy). Cities, 31, 349–356.

Salvati, L., & Carlucci, M. (2011). The economic and environmental performances of rural districts in Italy: Are competitiveness and sustainability compatible targets? Ecological Economics, 70(12), 2446–2453.

Salvati, L., & Carlucci, M. (2014). A composite index of sustainable development at the local scale: Italy as a case study. Ecological Indicators, 43, 162–171.

Salvati, L., & Zitti, M. (2009). Substitutability and weighting of ecological and economic indicators: Exploring the importance of various components of a synthetic index. Ecological Economics, 68(4), 1093–1099.

Salvati, L., Morelli, V. G., Rontos, K., & Sabbi, A. (2013). Latent exurban development: City expansion along the rural-to-urban gradient in growing and declining regions of southern Europe. Urban Geography, 34(3), 376–394.

Salvati, L., Venanzoni, G., Serra, P., & Carlucci, M. (2016). Scattered or polycentric? Untangling urban growth in three southern European metropolitan regions through exploratory spatial data analysis. The Annals of Regional Science, 57, 1–29.

Scheuer, S., Haase, D., & Volk, M. (2016). On the Nexus of the Spatial Dynamics of Global Urbanization and the Age of the City. Plos One, 11(8), e0160471.

Seifollahi-Aghmiuni, S., Kalantari, Z., Egidi, G., Gaburova, L., & Salvati, L. (2022). Urbanisation-driven land degradation and socioeconomic challenges in peri-urban areas: Insights from Southern Europe. Ambio, 51(6), 1446–1458.

Serra, P., Pons, X., & Saurí, D. (2008). Land-cover and land-use change in a Mediterranean landscape: a spatial analysis of driving forces integrating biophysical and human factors. Applied Geography, 28(3), 189–209.

Taylor, P.J., Firth, A., Hoyler, M., & Smith, D. (2010). Explosive city growth in the modern world-system: An initial inventory derived from urban demographic changes. Urban Geography, 31(7), 865–884.

Tomao, A., Quatrini, V., Corona, P., Ferrara, A., Lafortezza, R., & Salvati, L. (2017). Resilient landscapes in Mediterranean urban areas: Understanding factors influencing forest trends. Environmental Research, 156, 1–9.

Index

About the Editors

Alessia D'Agata is a visiting scholar at the Department of Geography, Autonomous University of Barcelona, and a doctoral candidate at the Department of Methods and Models for Territory, Economy and Finance, Faculty of Economics, Sapienza University of Rome, Italy, supported by a prestigious fellowship. She holds a master's in building engineering and architecture and a master's in GIScience for the integrated management of land and natural resources. Her main research interests include urban planning and policy, agricultural and environmental sustainability, official statistics, and normative approaches to sustainable development. She published articles in top-field scientific journals dealing with land-use change and forest dynamics.

Ioannis Konaxis, Ph.D., is an assistant professor of sustainable eco-tourism and nature conservation at the Business School of the public University of Piraeus, Greece. He teaches several classes on tourism business and economics, ecological conservation of protected areas, landscape, and sustainable development. He published theoretical and applied research in sustainable tourism management in prestigious international journals and book series.

Luca Salvati, Ph.D., is a senior researcher and professor of economic statistics at the Faculty of Economics, Sapienza University of Rome, Italy. His main research interests include regional and urban economics, housing policy, spatial planning, sustainable development, and indication theory. He has published more than 700 scientific articles in international journals and technical books on economic and statistical issues.